PROTEIN-PROTEIN INTERACTION ASSAYS

Edited by **Mahmood-ur-Rahman Ansari**

Protein-Protein Interaction Assays

http://dx.doi.org/10.5772/intechopen.71741

Edited by Mahmood-ur-Rahman Ansari

Contributors

Paolo Arcari, Immacolata Ruggiero, Nicola Martucci, Carmen Sanges, Emilia Rippa, Annalisa Lamberti, Ferdinando Papale, Vincenzo Quagliariello, Nunzia Migliaccio, Hiroshi Ueda, Yuki Ohmuro-Matsuyama, Manuel Fuentes, Kausar Malik, Haleema Sadia, Muhammad Hamza Basit, Mahmood-Ur-Rahman Ansari, Munazza Ijaz, Muhammad Iqbal

Notice

Statements and opinions expressed in the chapters are these of the individual contributors and not necessarily those of the editors or publisher. No responsibility is accepted for the accuracy of information contained in the published chapters. The publisher assumes no responsibility for any damage or injury to persons or property arising out of the use of any materials, instructions, methods or ideas contained in the book.

First published in London, United Kingdom, 2018 by IntechOpen

IntechOpen is the global imprint of INTECHOPEN LIMITED, registered in England and Wales, registration number: 11086078, The Shard, 25th floor, 32 London Bridge Street

London, SE19SG – United Kingdom

Printed in Croatia

British Library Cataloguing-in-Publication Data

A catalogue record for this book is available from the British Library

Additional hard copies can be obtained from orders@intechopen.com

Protein-Protein Interaction Assays, Edited by Mahmood-ur-Rahman Ansari

p. cm.

Print ISBN 978-1-78923-390-2

Online ISBN 978-1-78923-391-9

We are IntechOpen,
the world's leading publisher of
Open Access books
Built by scientists, for scientists

3,600+
Open access books available

113,000+
International authors and editors

115M+
Downloads

151
Countries delivered to

Our authors are among the

Top 1%
most cited scientists

12.2%
Contributors from top 500 universities

Interested in publishing with us?
Contact book.department@intechopen.com

Numbers displayed above are based on latest data collected.
For more information visit www.intechopen.com

Meet the editor

Mahmood-ur-Rahman Ansari received his BS degree (2003) in Plant Breeding and Genetics from the University of Agriculture, Faisalabad, Pakistan, and MPhil Degree (2006) and PhD Degree (2011) in Molecular Biology from National Centre of Excellence in Molecular Biology, University of the Punjab, Lahore, Pakistan. He has been a visiting fellow for 1 year in National Institute of Deafness and Other Communication Disorders, National Institutes of Health, USA. Currently, he is an assistant professor of Molecular Biology in the Department of Bioinformatics and Biotechnology, GC University, Faisalabad, Pakistan. He is also working as in charge of Molecular Biology Section, Central Hi-Tech Laboratory of the University. He has published over 40 papers in international peer-reviewed journals in the field of Molecular Biology, Biotechnology, and Bioinformatics. He is a member of various national and international professional societies and editorial board member/reviewer of numerous peer-reviewed journals. He is a founding member and member of Board of Directors of the Pakistan Society for Computational Biology (PSCB) since 2012. He also served as the General Secretary of PSCB from 2014 to 2016 for one term. He has been conducting research to understand the molecular mechanisms of stress tolerance in plants. He is interested to study protein-protein interactions of stress-tolerant proteins.

Contents

Preface

Protein-protein interactions (PPIs) are the physical contacts between two or more protein molecules as a result of biochemical reaction. These physical contacts or interactions occur in the cell or in the living organism for a specific biochemical event. The interaction between proteins shows the closeness of proteins for a specific function. Large number of related proteins closely works together in a living system to perform a specific task. PPIs are studied for different perspectives including biochemistry, chemistry, molecular dynamics, signal transduction, transport across membranes, cellular metabolism, etc. This book contains five chapters including an **introductory chapter**. This chapter deals with the introduction of protein-protein interactions and information about various assays to study PPIs.

Chapter 2 describes a novel PPI assay. The authors developed a novel assay named FlimPIA using two mutant Flucs, each of which catalyzes one of the two half reactions catalyzed by the wild-type enzyme. It shows many advantages over other assays, such as longer detectable distance, more stable probes, and higher signal readout in a shorter time period. **Chapter 3** discusses various applications of nanomaterials in PPIs. In this chapter, several critical aspects related to the protein corona are described. **Chapter 4** highlights different protein-based methods to detect genetically modified organisms (GMOs) in the environment. The development of GM crops is rapidly expanding every year around the world. So, it is the need of the hour to develop suitable methods for their detection. Different immunoassays or catalyst connected immunosorbent tests discussed here are sensitive and more affordable. The authors also developed a strip assay to detect Bt genes in GM plants. **Chapter 5** describes the interactions of eukaryotic translation elongation factor 1A (eEF1A), which is involved in many cellular processes such as cell survival and apoptosis. The authors showed that eEF1A phosphorylation occurred only in the presence of eEF1A1 and eEF1A2, thus suggesting that both isoforms interacted in cancer cells.

The valuable information available in these chapters will advance the knowledge of research students, researchers, academician and general public who are interested in this topic. At the end, I thank the **IntechOpen Book Department** for giving me an opportunity to edit this book. I am also very much thankful to **Ms. Danijela Sakic**, Publication Manager, for her valuable help throughout the editing process. I must thank my research student **Ms. Munazza Ijaz** for her assistance in handling the chapters. Many thanks are given to all authors for their precious contributions.

Mahmood-ur-Rahman Ansari, PhD
Department of Bioinformatics & Biotechnology
GC University, Faisalabad
Pakistan

Introduction

Chapter 1

Introductory Chapter: Protein-Protein Interactions and Assays

Munazza Ijaz, Mahmood-ur-Rahman Ansari and
Muhammad Iqbal

Additional information is available at the end of the chapter

http://dx.doi.org/10.5772/intechopen.77337

1. Introduction

Protein-protein interactions (PPIs) control variety of biological phenomena including development, cell to cell interactions and metabolic processes [1]. The PPIs can be classified into different groups, depending upon their functional and structural properties [2]. Depending upon their persistence, (1) they may be termed as permanent or transient, as characterized by their interaction surface, (2) they may be considered as heterooligomeric or homooligomeric based on their stability, and (3) they may be called as obligate or nonobligate [3]. A blend of these three pairs may develop a protein-protein interaction. For example, a permanent interaction of the protein may be able to form a stable protein complex while on the other hand a transient interaction among the proteins may form a signaling pathway [4].

To perform the function in a living cell, proteins rarely act as isolated species [5] instead over 80% of the proteins perform their functions in groups [6]. The function of an unidentified protein can be suggested by its interactions with a protein of known function. The thorough study of PPIs is also important to demonstrate the molecular mechanism of cellular processes of proteins [3]. The momentous properties of PPIs are (a) the kinetic properties of the enzymes can be modified by PPIs; (b) PPIs can allow substrate channeling; (c) they can create a new binding site for the small molecules; (d) PPIs can suppress or activate a protein; (e) PPIs can perform regulatory role in downstream or upstream regulation of the protein; and (f) they can also alter the specificity of binding of the protein for its particular substrate by changing its interactions [7].

The proteins that have many interactions include transcription factors and enzymes [8]. Though, PPIs encompass heterogeneous procedures mostly and the possibility of their regulation is enormous. Various interactions and the outcome of these interactions are needed to identify the better understanding of PPIs inside the cell [3]. By using methods like mass spectrometry, protein chip technology, phage display, and two hybrid system, PPIs data have been increased in recent years [3]. These experimental resources are useful for constructing comprehensive PPI networks. But, day by day the increase in the amount of data on protein-protein interactions is becoming a challenge for validation in the laboratory. To understand the functions of unexplored protein by using computational approaches is necessary nowadays.

2. Protein-protein interaction assays

Protein-protein interaction (PPI) assays can be classified into three broad categories, i.e., in vivo, in vitro, and in silico. (1) The in vivo techniques apply the whole procedure on the living cell or organism itself. (2) In vitro methods require the whole procedure completed outside the cell in a controlled environment in a laboratory, i.e., affinity chromatography, tandem affinity purification (TAP), protein fragment complementation, X-ray crystallography, co-immunoprecipitation, phage display, nuclear magnetic resonance, spectroscopy, and protein arrays. (3) The techniques that are performed by using computers or computer simulations are called in silico techniques. The sequence and structure-based approaches, gene fusion, chromosome proximity, gene expression-based approaches, domain pair-based approach, in silico two hybrid approaches, phylogenetic profile, and phylogenetic tree are some approaches which are based on in silico methods [4].

2.1. In vivo techniques for the prediction of protein-protein interactions

The in vivo technique used to study PPIs is yeast two hybrid (Y2H) method [9]. The two protein domains are involved in the Y2H assay. First domain is the DNA binding domain which helps in binding the DNA and the second one is activation domain that is involved in activation of the transcription of the specific DNA. These two domains are required for the transcription of a particular reporter gene [10]. The interacting proteins that are involved in the Y2H assays must be present in the close vicinity or inside the nucleus because these proteins have the capability to activate reporter gene and the proteins that are not present in nucleus do not have the ability to activate reporter genes. Some other techniques being used are fluorescence resonance energy transfer (FRET), biomolecular fluorescence complementation (BiFC), and bioluminescence resonance energy transfer (BRET) [4].

2.2. In vitro techniques for the prediction of protein-protein interactions

To learn PPIs in the inherent environment of the cell, a technique called TAP tagging [11] was developed. TAP tagging method was first used to analyze the yeast interactome in a high throughput way [12]. TAP tagging involves two steps, first is double tagging of the protein of interest and second is two-step process of purification [13]. After the process, the proteins that remain attached with the target protein can be studied by using SDS-PAGE and then mass spectrometry analysis is performed to confirm the PPI partner of the protein of interest [14]. TAP tagging used in combination with mass spectrometry which can identify both protein complexes and protein interactions.

Affinity chromatography is also used to study PPIs in vitro. It is very sensitive technique and can identify even the weakest interactions among the proteins. Though, it generates many of the false positive results because of the great specificity of the proteins. Therefore, studies of protein interactions cannot depend only on affinity chromatography. So, other techniques are needed in combination with affinity chromatography to further confirm the results generated. The affinity chromatography is often combined with mass spectroscopy and SDS-PAGE to produce more convincing results [4].

Co-immunoprecipitation is another in vitro technique that is used for the confirmation of PPIs by using the complete cell extract wherever proteins are present in a complicated blend of cellular machinery and in their natural form that is essential for the significant interactions of proteins [4]. Protein arrays are also being used to study PPIs. A piece of glass is used in which different protein molecules are attached in an organized fashion [15]. Protein microarray analysis gained marvelous importance to do high throughput analysis by running many of the parallel analysis in an automated procedure. The PPIs can be studied by using another proteomics method known as protein fragment complementation assay (PCA). It consists of a family of assays that can be used to identify the proteins of any molecular weight and it provides very simple and straight conducts to determine PPIs in living cells, in vitro, and multicellular organisms [4].

Mass spectroscopy can also be used to determine protein-protein interactions. There are two ways to identify PPIs by using mass spectroscopy shotgun proteomics and peptide finger printing [16]. To analyze complicated mixtures, shotgun proteomics is the most suitable technique while in the peptide finger printing, SDS-PAGE is used to separate the eluted complex. X-ray crystallography can also be used to determine PPIs in vitro. It is a type of microscopy with very high resolution that is used for the identification of proteins at atomic level and it is helpful for functional analysis of proteins [17]. The analysis of PPIs can also be done by using nuclear magnetic resonance (NMR) spectroscopy. In NMR spectroscopy, the magnetically active nuclei that are surrounded by a strong magnetic field engross electromagnetic radiations at distinguishing frequencies that are monitored by the chemical surroundings [18].

2.3. In silico techniques to predict protein-protein interactions

Many of the in vivo and in vitro techniques generate a large amount of data that is helpful in the development of software and tools for the identification of PPIs among various proteins that are found in many different combinations. The computational methods used for the in silico prediction of interactions among proteins may include the tools described in **Table 1**.

S. No.	Tool name	Link
1	Coev2Net	http://groups.csail.mit.edu/cb/coev2net/
2	TSEMA	http://tsema.bioinfo.cnio.es/
3	InterPreTS	http://www.russell.embl.de/interprets
4	Struct2Net	http://groups.csail.mit.edu/cb/struct2net/webserver/
5	PoiNet	http://poinet.bioinformatics.tw/
6	PrePPI	http://bhapp.c2b2.columbia.edu/PrePPI/
7	iWARP	http://groups.csail.mit.edu/cb/iwrap/
8	PIPE2	http://cgmlab.carleton.ca/PIPE2
9	PreSPI	http://code.google.com/p/prespi/
10	SPPS	http://mdl.shsmu.edu.cn/SPPS/
11	HomoMINT	http://mint.bio.uniroma2.it/HomoMINT
12	P-POD	http://ppod.princeton.edu/
13	BLASTO	http://oxytricha.princeton.edu/BlastO/
14	PHOG	http://phylogenomics.berkeley.edu/phog/
15	COG	http://www.ncbi.nlm.nih.gov/COG/
16	OrthoMCL-DB	http://orthomcl.org/orthomcl/
17	STRING	http://string.embl.de
18	MirrorTree	http://csbg.cnb.csic.es/mtserver/
19	G-NEST	https://github.com/dglemay/G-NEST
20	InPrePPI	http://inpreppi.biosino.org/InPrePPI/index.jsp
21	PRISM PROTOCOL	http://prism.ccbb.ku.edu.tr/prismprotocol/

Table 1. The list of in silico tools and their links to predict protein – protein interactions.

Acknowledgements

This work was supported by funds from Higher Education Commission of Pakistan.

Authors' contributions

MI did the research work and wrote first draft of manuscript, MR designed and wrote the paper and MI assisted in writing the paper. All the authors read the manuscript and approved for publication.

Author details

Munazza Ijaz[1], Mahmood-ur-Rahman Ansari[1,2]* and Muhammad Iqbal[3]

*Address all correspondence to: mahmoodansari@gcuf.edu.pk

1 Department of Bioinformatics and Biotechnology, GC University-Faisalabad, Faisalabad, Pakistan

2 Central Hi-Tech Lab, GC University-Faisalabad, Faisalabad, Pakistan

3 Department of Environmental Science and Engineering, GC University-Faisalabad, Faisalabad, Pakistan

References

[1] Braun P, Gingras AC. History of protein-protein interactions: From egg-white to complex networks. Proteomics. 2012;**12**(10):1478-1498. DOI: 10.1002/pmic.201100563

[2] Nooren IMA, Thornton JM. Diversity of protein-protein interactions. The EMBO Journal. 2003;**22**(14):3486-3492. DOI: 10.1093/emboj/cdg359

[3] Zhang A. Protein Interaction Networks: Computational Analysis. New York, NY: Cambridge University Press; 2009

[4] Rao VS, Srinivas K, Sujini GN, Kumar GNS. Protein-protein interaction detection: Methods and analysis. International Journal of Proteomics. 2014;**2014**(2):1-12. DOI: 10.1155/2014/147648

[5] Yanagida M. Functional proteomics; current achievements. Journal of Chromatography. B, Analytical Technologies in the Biomedical and Life Sciences. 2002;**771**(1-2):89-106. DOI: 10.1016/S1570-0232(02)00074-0

[6] Berggård T, Linse S, James P. Methods for the detection and analysis of protein-protein interactions. Proteomics. 2007;**7**(16):2833-2842. DOI: 10.1002/pmic.200700131

[7] Phizicky EM, Fields S. Protein-protein interactions: Methods for detection and analysis. Microbiological Reviews. 1995;**59**(1):94-123. DOI: 10.1111/j.1471-4159.2009.06024.x

[8] Sarmady M, Dampier W, Tozeren A. HIV protein sequence hotspots for crosstalk with host hub proteins. PLoS One. 2011;**6**(8). DOI: 10.1371/journal.pone.0023293

[9] Uetz P, Giot L, Cagney G, Mansfield TA, Judson RS, Knight JR, Lockshon D, Narayan V, Srinivasan M, Pochart P, Qureshi-Emili A, Li Y, Godwin B, Conover D, Kalbfleisch T, Vijayadamodar G, Yang M, Johnston M, Fields S, Rothberg JMA. Comprehensive analysis of protein-protein interactions in Saccharomyces cerevisiae. Nature. 2000;**403**(6770):623-627. DOI: 10.1038/35001009

[10] Ito T, Chiba T, Ozawa R, Yoshida M, Hattori M, Sakaki Y. A comprehensive two-hybrid analysis to explore the yeast protein interactome. Proceedings of the National Academy of Sciences of the United States of America. 2001;**98**(8):4569-4574. DOI: 10.1073/pnas.061034498

[11] Rigaut G, Shevchenko A, Rutz B, Wilm M, Mann M, Seraphin B. A generic protein purification method for protein complex characterization and proteome exploration. Nature Biotechnology. 1999;**17**(10):1030-1032. DOI: 10.1038/13732

[12] Gavin AC, Bösche M, Krause R, Grandi P, Marzioch M, Bauer A, Schultz J, Rick JM, Michon AM, Cruciat CM, Remor M, Höfert C, Schelder M, Brajenovic M, Ruffner H, Merino A, Klein K, Hudak M, Dickson D, Rudi T, Gnau V, Bauch A, Bastuck S, Huhse B, Leutwein C, Heurtier MA, Copley RR, Edelmann A, Querfurth E, Rybin V, Drewes G, Raida M, Bouwmeester T, Bork P, Seraphin B, Kuster B, Neubauer G, Superti-Furga G. Functional organization of the yeast proteome by systematic analysis of protein complexes. Nature. 2002;**10, 415**(6868):141-147

[13] Pitre S, James A, Michel RG. Computational methods for predicting protein-protein interactions. Advances in Biochemical Engineering/Biotechnology. 2008;**110**:247-267. DOI: 10.1007/10

[14] Rohila JS, Chen M, Cerny R, Fromm ME. Improved tandem affinity purification tag and methods for isolation of protein heterocomplexes from plants. The Plant Journal. 2004;**38**(1):172-181. DOI: 10.1111/j.1365-313X.2004.02031.x

[15] MacBeath G, Schreiber SL. Printing proteins as microarrays for high-throughput function determination. Science. 2000;**289**:1760-1763. DOI: 10.1126/science.289.5485.1760

[16] Moresco JJ, Carvalho PC, Yates JR. Identifying components of protein complexes in *C. elegans* using co-immunoprecipitation and mass spectrometry. Journal of Proteomics. 2010;**73**:2198-2204. DOI: 10.1016/j.jprot.2010.05.008

[17] Tong AHY, Evangelista M, Parsons AB, Xu H, Bader GD, Pagé N, Robinson M, Raghibizadeh S, Hogue CW, Bussey H, Andrews B, Tyers M, Boone C. Systematic genetic analysis with ordered arrays of yeast deletion mutants. Science. 2001;**294**(5550):2364-2368. DOI: 10.1126/science.1065810

[18] O'Connell MR, Gamsjaeger R, Mackay JP. The structural analysis of protein-protein interactions by NMR spectroscopy. Proteomics. 2009;**9**(23):5224-5232. DOI: 10.1002/pmic.200900303

Protein-Protein Interaction Assays

A Novel Protein-Protein Interaction Assay Based on the Functional Complementation of Mutant Firefly Luciferases: Split Structure Versus Divided Reaction

Yuki Ohmuro-Matsuyama and Hiroshi Ueda

Additional information is available at the end of the chapter

http://dx.doi.org/10.5772/intechopen.75644

Abstract

Protein-fragment complementation assays (PCAs) are commonly used to assay protein–protein interaction (PPI). While PCAs based on firefly luciferase (Fluc) in cells or lysates are a user-friendly method giving a high signal/background (S/B) ratio, they are difficult to use in vitro owing to the instability of split Fluc fragments. As a solution to this issue, we developed a novel protein–protein interaction assay named FlimPIA using two mutant Flucs, each of which catalyzes one of the two half-reactions catalyzed by the wild-type enzyme. Upon approximation by the tethered protein pairs, the two mutants yielded higher signal owing to a more efficient transfer of the reaction intermediate luciferyl adenylate. FlimPIA showed many advantages over in vitro split Fluc assays, such as longer detectable distance, more stable probes, and higher signal readout in a shorter time period, and it also worked in cellulo.

Keywords: protein–protein interaction assay, firefly luciferase, protein-fragment complementation assay, FlimPIA, FRET

1. Introduction

When the human genome project was completed in 2003, most researchers expected dramatic developments in various fields such as biology, etiology, and drug discovery. However, progression did not remarkably accelerate. One of the causes is that protein-protein interactions (PPIs) are still not well understood. In the cell, many proteins interact with each other and cooperate to fulfill their roles in biological phenomena. It is reported that there are 150,000–300,000 PPIs in the human interactome [1, 2]. Therefore, PPI assays are very important for biology, diagnosis, and drug discovery.

The conventional PPI assays, which are available both in vitro and in cellulo, are Förster/fluorescence resonance energy transfer (FRET)-based assays, bioluminescent resonance energy transfer (BRET) assays, and protein-fragment complementation assays (PCAs).

For FRET-based assays, two fluorescent proteins or two fluorescent dyes are fused to proteins that interact with each other. When the interaction occurs, the two fluorescent proteins (dyes) are in close proximity, and then the energy transfer is induced, resulting in changes of the fluorescent intensities. In BRET assays, a bioluminescent enzyme and fluorescent protein (dye) are fused to proteins that interact with each other, and the energy is transferred from the bioluminescent enzyme to the fluorescent protein (dye). FRET- and BRET-based assays are the most common and sophisticated methods.

For PCA, the enzyme or fluorescence is divided into two fragments. The split fragments are fused to interacting proteins. Upon interaction, the split fragments come close, and then the full length of the structure is reconstituted, resulting in the recovery of the enzyme activity or fluorescence. PCA in cells and lysates is a user-friendly method that gives a high signal/background (S/B) ratio [3]. Moreover, we reported in vitro PCA using purified firefly luciferase (Fluc) fragments for the first time [4, 5]. The development of PCA is described in Section 2.

Recently, we developed a novel PPI assay, named firefly luminescent intermediate-based protein-protein interaction assay, FlimPIA [6–10]. FlimPIA utilizes the unique reaction of Fluc, which is divided into two half steps. We describe the principle of FlimPIA in Section 3.1 and the several improvements of FlimPIA in Sections 3.2–3.6. Then, the advantages and disadvantages of FlimPIA compared to another PPI assays such as the in vitro PCA are described in the final section.

2. Demonstration of in vitro protein-fragment complementation assay using purified Fluc fragments

Conventional PCA is used in vivo and in cultured cells (in cellulo). Although Porter et al. performed a Fluc-based PCA in vitro, the assay requires cell lysate, and the components in the lysate might affect the PPIs. We succeeded in developing a Fluc-based PCA in vitro using purified probes in a defined solution [4, 5].

For the Fluc-based PCA in vitro, a well-known interacting pair, FKBP12 (a 12 kD domain of FK506-binding protein) and FRB (FKBP-rapamycin-associated protein), was utilized. The association between these proteins depends on the presence of an antibiotic, rapamycin [11, 12]. Two pairs of split *Photinus pyralis* Fluc—the pair of the N-terminal domain (amino acids [aa] 1–437) and the C-terminal domain (aa 394–547) and the pair of the N-terminal domain (aa 1–398) and the same C-terminal domain (aa 394–547)—were selected in several split sites of Fluc [13], which worked well for in cellulo PCA. The gene encoding FKBP12 or FRB was fused to the 5′ end of each domain, and the genes were inserted into the pET32 vector, which originally encodes thioredoxin (Trx), yielding four fusion protein genes, FKBP-N, FKBP-C, FRB-N, and FRB-C. These proteins were expressed in the soluble fraction of *E. coli* BL21(DE3) pLysS and purified by immobilized metal affinity chromatography (**Figure 1A**).

The two interacting pairs, FKBP-N and FRB-C, FKBP-C and FRB-N, were mixed, and rapamycin was added to the pair (**Figure 1B**). The luminescence intensities of the mixture of the interacting pairs and rapamycin were remarkably increased immediately after adding the two

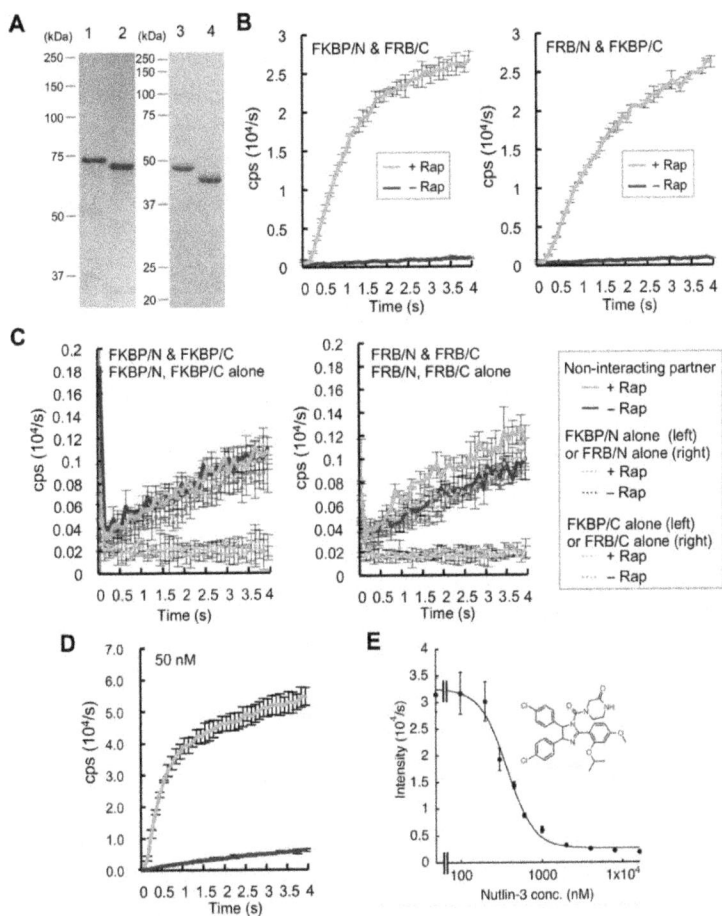

Figure 1. Detection via Fluc-based PCA using purified probes. (A–C) Detection of FKBP-FRB association. (A) Purified probes. Lane 1, FKBP-N; Lane 2, FRB-N; Lane 3, FRB-N; Lane 4, FRB-C. (B) PCA using the purified probes at 50 nM each, with/without equimolar rapamycin (n = 3). (C) Control experiments (n = 3). (D, E) Detection of p53-Mdm2 association. (D) PCA using the purified probes at 50 nM each (n = 3). (E) Inhibition of p53-Mdm2 interaction by Nutlin-3 (n = 3). ©American Chemical Society

substrates, luciferin and ATP. On the other hand, the luminescence intensities of the mixture of the interacting pair (FKBP12 and FRB) without rapamycin and noninteracting pair were very low (**Figure 1C**). The signal and stability of the pair of the N-terminal domain (aa 1–437) and the C-terminal domain (aa 384–547) were higher than those of the other pair of N-terminal domain (aa 1–398) and the same C-terminal domain. When the first pair was used, the luminescence signal displayed rapamycin dose dependence, and the limit of detection was determined as 250 pM. These results clearly showed that the PPI could be detected with a high S/B ratio and high sensitivity using the purified probes.

Because the rapamycin-dependent FKBP-FRB association is very strong, another interacting pair, p53 and Mdm2, was investigated (**Figure 1D** and **E**) [14, 15]. p53 suppresses cell growth as a tumor suppressor. The oncoprotein Mdm2 binds to p53 and downregulates the function of p53 in certain cancer cells. In the assay of p53-Mdm2 interaction using p53-C and Mdm2-N, the signal intensity and S/B ratio rose with higher concentrations of the probes of the interacting pair. To investigate the reversibility of the PCA, an inhibitor of the p53-Mdm2 interaction, Nutlin-3, was added to the mixture of p53-C and Mdm2-N. The luminescence intensity decreased depending on the concentration of Nutlin-3.

The in vitro PCA opens the way to study PPIs of cytotoxic proteins, which is impossible to perform in cells. Furthermore, the possibility that the cellular components affect PPIs can be excluded.

3. Development of a novel PPI assay FlimPIA

In this section, we describe a novel PPI assay, FlimPIA, which we recently developed and continue to improve.

3.1. Principle of FlimPIA

In contrast to PCA, in which the structure of Fluc is divided into two domains as the probes, FlimPIA divides the reaction catalyzed by Fluc into two half-reactions. Fluc catalyzes the conversion of firefly D-luciferin (LH_2) to the excited state oxyluciferin (OxL) by a two-step catalysis, namely, an adenylation step and oxidative luminescence steps. In the adenylation step, LH_2 is converted to D-luciferyl adenylate (LH_2-AMP), and in the oxidative luminescence steps, LH_2-AMP is converted to OxL, and then excited OxL emits light. It was recently supposed that Fluc, which consists of a large N-terminal domain and a small C-terminal domain connected by a flexible hinge region, rotates its C-terminal domain by ~140° to proceed from the adenylation step to the oxidative luminescence steps (**Figure 2**) [16, 17]. One reason for this hypothesis is that the active site of each step in acyl-adenylate-forming enzymes, including Fluc, is different. In the adenylation step, K529 is an important amino acid residue, and on the other hand, K443 and H245 are key residues for the oxidative luminescence steps [18–20].

Two mutant *Photinus pyralis* Flucs were designed for FlimPIA; one is H245D/K443A/L530R, which can produce LH_2-AMP but cannot catalyze LH_2-AMP to form OxL, and the other is K529Q, which very slowly produces LH_2-AMP but maintains the catalytic steps in the oxidative luminescence half-reaction. Each mutant is fused to proteins that interact with each other. The interaction brings the mutants close together, and then LH_2-AMP, which H245D/K443A/L530R produces, is utilized by K529Q, resulting in OxL production (**Figure 3**). The mutant H245D/K443A/L530R acts as the "Donor" providing LH_2-AMP, and the mutant K529Q works as the "Acceptor" of LH_2-AMP [7].

When FKBP12 and FRB are fused to the Donor and Acceptor, respectively, the luminescence intensity increased depending on the concentration of rapamycin (**Figure 4A, B**). The EC_{50} values of the cognate pairs were 10.2 ± 0.6 and 16.0 ± 2.1 nM, respectively, which correspond well with the reported K_D value of the association between FKBP12/rapamycin and FRB. FK506 (tacrolimus) is commonly used as an immunosuppressant to prevent the rejection of organ transplants and inhibits the rapamycin-dependent association between FKBP12 and FRB [14]. The luminescence intensity decreased upon FK506 addition (**Figure 4C**). The S/B ratio increased depending on the concentration of PPI when the concentration of probes and

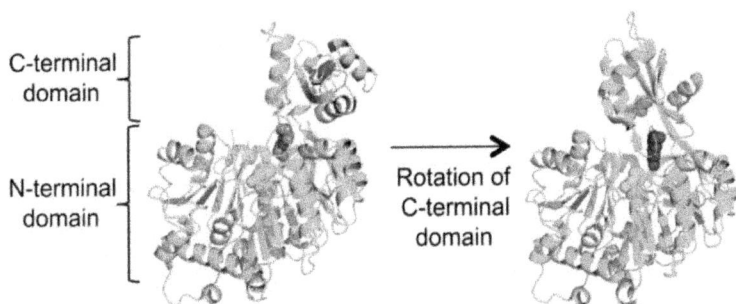

Figure 2. The conformational change of Fluc. Fluc is composed of a large N-terminal domain and a smaller C-terminal domain, which rotates ~140° according to the reactions to proceed from the adenylation reaction to the oxidative luminescent reactions: Key Lys residues (K529 and K443) are shown in light and dark blue, respectively. Another key residue H245 is shown in cyan. ©American Chemical Society.

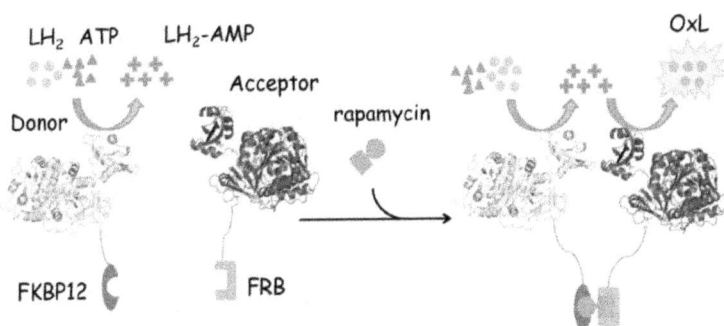

Figure 3. The working principle of FlimPIA. ©American Chemical Society.

rapamycin was up to 500 nM (**Figure 4D**). In addition, the association between FKBP12 and FRB could be detected in 40% fetal bovine serum diluted in phosphate buffered saline, suggesting the applicability of the assay to clinical samples.

Next, p53 and Mdm2 were used as interacting proteins. The luminescence intensity of the mixture of the interacting pair (p53-Donor and Mdm2-Acceptor) was higher than the intensities of noninteracting pairs (p53-Donor and p53-Acceptor, Mdm2-Donor and Mdm2-Acceptor) (**Figure 4E**). The inhibition of the p53-Mdm2 interaction by Nutlin-3 was observed (**Figure 4F**). The result clearly shows that FlimPIA is a versatile system and can analyze transient interactions.

3.2. Improved FlimPIA by the entrapment of Fluc conformation

The original FlimPIA had exhibited high background signal, which was mainly caused by the remaining adenylation activity of the Acceptor. As mentioned above, the C-terminal domain rotates according to the reactions proceeding from the adenylation to the oxidative luminescence reactions (**Figure 2**). Therefore, we tried to entrap the Acceptor conformation into the oxidation conformation [10].

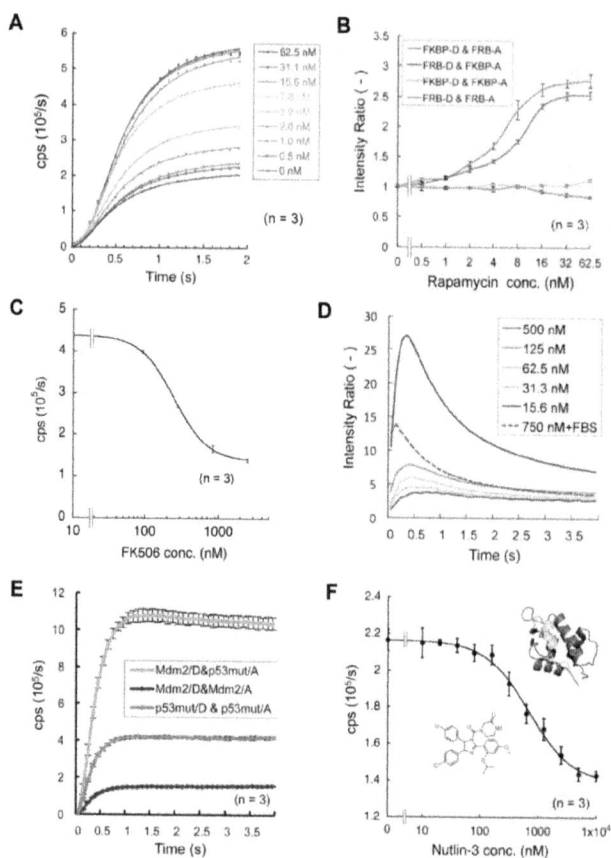

Figure 4. Detection via FlimPIA in vitro. (A-D) Detection of FKBP-FRB association. (A) Luminescence time course at several rapamycin concentrations. A mixture of FKBP/Donor and FRB/Acceptor (50 nM each) was used. (n = 3). (B) Specific detection of FKBP12-FRB interaction. The four possible combinations of four Fluc mutants, namely, FKBP/Donor, FRB/Donor, FKBP/Acceptor and FRB/Acceptor (50 nM each) were tested for their rapamycin dose-dependency. The relative luminescence integrated for 1.5–1.6 s after substrate addition is shown (n = 3). (C) Competition of PPI (protein–protein interaction) by FK506. Rapamycin (80 nM) and FK506 at indicated concentration were added to the mixture of FKBP/Donor and FRB/Acceptor (80 nM each). The luminescence integrated for 0.8–0.9 s after substrate addition is shown (n = 3). (D) Time course of S/B (signal/background) ratio obtained with the mixture of FKBP/Donor and FRB/Acceptor with and without equimolar rapamycin. The ratio of the two light intensities at the indicated time point is shown. Sample with 40% fetal bovine serum and 750 nM proteins is also shown (n = 3). (E–F) Detection of p53-Mdm2 association. (E) Luminescence time course of the cognate (Mdm2/Donor and p53/Acceptor) and control pairs (25 nM each) (n = 3). (F) Competition of PPI by a specific inhibitor. Nutlin-3 (bottom) at indicated concentration was added to the mixture of p53/Donor and Mdm2/Acceptor (25 nM each) (n = 3). The luminescence integrated for 0.8–0.9 s after substrate addition is shown. The ribbon model of Mdm2 (purple)-p53 peptide (light green) complex is also shown. ©American Chemical Society.

According to the report by Branchini et al. that the structure of Fluc could be fixed into the oxidative luminescence conformation by chemical trapping, we first took the same approach to entrap Acceptor mutant [21]. Specifically, all cysteine residues in the Acceptor were substituted with serine or alanine residues. Then, the residues at positions 108 and 447 were substituted with cysteine residues and cross-linked by 1,2-bis-(maleimide)ethane (BMOE) (**Figure 5A**).

The luminescence of the cross-linked Acceptor was almost diminished compared to the non-cross-linked Acceptor (**Figure 5B**). The Acceptor and Donor were fused to FRB and FKBP12, respectively. In a FlimPIA using the cross-linked Acceptor, the background signal was eliminated, and the signal induced by the interaction was significantly higher than the background signal (**Figure 5C**). Taken together, the results clearly showed that the Acceptor can be trapped into the oxidation conformation and the sensitive FlimPIA was successfully developed, giving a high S/B ratio.

As the substitution of the all cysteine residues considerably reduced the luminescence intensity, next, we tried to use the original Acceptor retaining the cysteine residues and put the cysteine residues at positions 108 and 447, which were then cross-linked by BMOE. As a result, one-fifth of the luminescence intensity of the cross-linked Acceptor was retained, probably due to miss- and/or incomplete cross-linking (**Figure 5D**). Although there was some background signal, an apparent improvement in luminescent intensity was observed. When the same concentration (50 nM each) of the probes and rapamycin were used, the maximum S/B ratio was improved from 2.6 to 5.3, compared with the original system (**Figure 5E**).

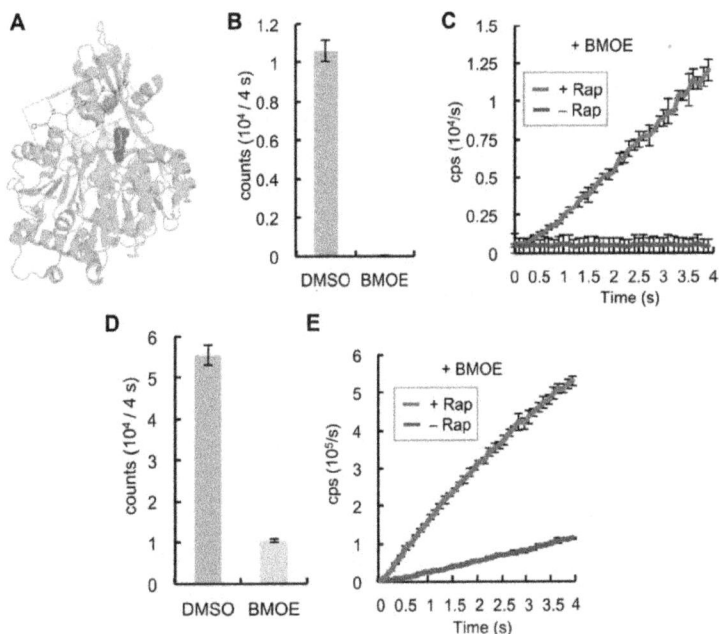

Figure 5. FlimPIA using the trapped Acceptor by bis-maleimide crosslinker (A–C) The trapping by bis-maleimide crosslinker (1). (D–E) The trapping by bis-maleimide crosslinker (2). (A) Scheme of the trapped Acceptors by BMOE. Residues shown in yellow were used for the N–C linkage. (B, D) Suppression of overall luminescent activity by chemical trapping of the Acceptor. The enzyme (10 nM) was reacted with 75 μM LH$_2$ and 10 mM ATP. The luminescent intensities with and without chemical modification by BMOE were compared (n = 3). (C, E) The mixture of FKBP/Donor and trapped FRB/Acceptor (50 nM each) was added with/without 50 nM rapamycin (n = 3). ©American Chemical Society.

3.3. Improved FlimPIA using a mutant acceptor (1)

During another attempt to select paired cysteine residues for possible cross-linking of N-C domains, the introduction of S198C/S440C mutations on the background of original Acceptor was attempted. However, the obtained clone was later found to be contaminated with the S440C mutant retaining only one mutation. The resultant S440C mutant showed higher ability as the Acceptor, whereas the S198C/S440C mutant did not act as the Acceptor. To understand the effect of this mutation, we performed saturation mutagenesis of the S440 residue. The substitution of leucine, phenylalanine, and tryptophan, which have bulky and/ or large side chains, gave a higher maximal S/B ratio in FlimPIA (**Table 1**) [9]. Additionally, not all the mutants with bulky or long side chains showed higher S/B ratios. Although the precise reason is not known, it might be because mutations often affect protein stability and/ or aggregation.

We expected that the bulky and/or large side chains at this position could form steric hindrance with hinge region and the C-terminal domain from the structural modeling based on the adenylation conformation structure of *Luciola cruciata* Fluc with bound substrate analog (**Figure 6A**). On the other hand, there seemed no severe inhibition in the model of the oxidative luminescence conformation.

Then we examined the adenylation and oxidative luminescence activities of the S440L Acceptor. The amounts of LH_2-AMP produced by the new and conventional Acceptors were examined according to the method using the N-terminal domain of Fluc as a selective detector of LH_2-AMP [18]. The LH_2-AMP produced by the new Acceptor was less than one-fifth of the LH_2-AMP produced by the conventional Acceptor (**Figure 6B**). On the other hand, the kinetics against LH_2-AMP are shown in **Table 2**. Because the concentration of the LH_2-AMP that the

S440X	S/B ratio	S440X	S/B ratio
L	7.93 ± 0.60	Q	2.11 ± 0.01
F	5.69 ± 0.12	R	2.08 ± 0.41
W	4.94 ± 0.06	S	1.87 ± 0.24
M	3.65 ± 0.35	N	1.86 ± 0.08
K	3.45 ± 0.20	V	1.85 ± 0.25
A	2.86 ± 0.22	D	1.80 ± 0.13
Y	2.81 ± 0.37	G	1.67 ± 0.26
H	2.57 ± 0.19	I	1.55 ± 0.21
C	2.32 ± 0.13	T	1.52 ± 0.22
E	2.31 ± 0.10	P	1.09 ± 0.11

Table 1. Comparison of maximum S/B ratios obtained by S440 mutants.

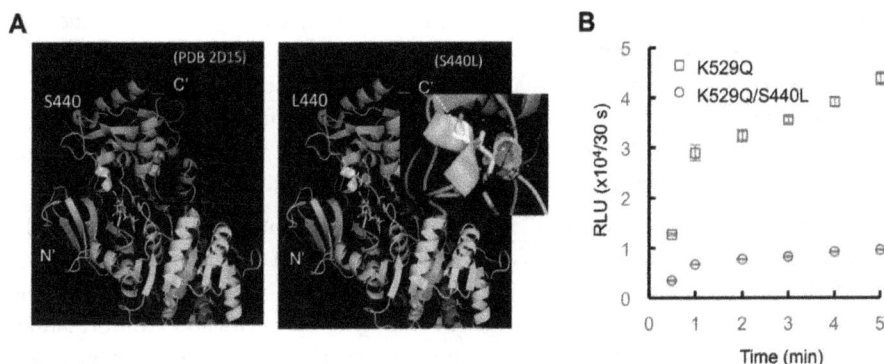

Figure 6. Possible steric hindrance of adenylation conformation with the S440 L mutation. (A) Structure of Fluc (left) and a model of Fluc S440L (right), each at adenylation conformation. The Leu440 residue (shown in white) is enlarged in the inset. Drawn with PyMOL software. (B) Adenylation activity measured by the N-terminal domain method. Error bars mean ±SD (n = 3).

	V_{max} ($\times 10^6$ RLU*/sec)	K_m (µM)	V_{max}/K_m ($\times 10^6$ RLU/s µM^{-1})
K529Q	1.40 ± 0.16	0.513 ± 0.018	2.11 ± 0.01
K529Q/S440L	0.296 ± 0.031	0.321 ± 0.013	2.08 ± 0.41

*Relative light units

Table 2. Oxidative luminescence activity of K529Q and S440L/K529Q (1 nM each).

Acceptor uses in FlimPIA is low, the V_{max}/K_m is the most important kinetics parameter. The value of the new Acceptor decreased to 33.6% of the value of the conventional Acceptor; therefore, the luminescence intensity in FlimPIA might decrease to some extent. Taken together, the balance of the adenylation and oxidative activities of the new Acceptor gave the highest S/B ratio in the Acceptors, which we have developed.

3.4. Improved FlimPIA using mutated acceptor (2)

When the C-terminal domain of Fluc rotated to proceed from the adenylation step to the oxidative luminescence steps, the flexible hinge region between N- and C-terminal domains is considered highly important (**Figure 2**). Furthermore, the hinge region sits close to the active site in the adenylation conformation. To obtain suitable mutants for the Acceptor, semi-random mutations at the residues 436–439 in the hinge region were introduced [6]. The amino acid residues that enzymes in acyl-adenylate-forming enzyme superfamily contain at the corresponding positions were chosen in the semi-random library. The mutant R437K/L438I was selected from the library, because the mutants showed lower adenylation activity (~15% of the wild-type Fluc) and slightly higher oxidative luminescence activity (116% of the wild-type Fluc).

A single mutation, R437K, or a double mutation, R437K/L438I, was introduced into the conventional Acceptor (K529Q). The overall luminescence activity and the oxidative luminescence activity of the two new Acceptors were compared to that of the conventional Acceptor (**Figure 7A, B**). The overall activities of both new Acceptors decreased almost tenfold compared with that of the conventional Acceptor, whereas the oxidative luminescence activities were almost maintained. These results showed that R437K is a key residue for Acceptor activity.

The kinetics properties of the conventional and the new Acceptors fused to FRB are shown in **Table 3**. The lower overall activities and the similar oxidative luminescence activities are probably due to the remarkably lower V_{max} values for LH_2 and ATP and similar V_{max} and K_m values for LH_2-AMP. Moreover, in the structural model of the adenylation conformation, the mutated residue K437 sits close to the active site residues such as K529, suggesting some inhibition of the adenylation activity (**Figure 7C**).

When the FKBP12-FRB interaction was detected by FlimPIA, the maximum S/B ratio reached approximately 4, whereas it was approximately 2.5 in the conventional assay (**Figure 7D**). Taken together, we succeeded in finding a suitable mutant for the Acceptor in the semirandom library of the hinge region. Furthermore, these results suggest that the hinge region is important for controlling the two half-reactions of Fluc and supports the hypothesis that the C-terminal domain rotates to accomplish the half-reactions.

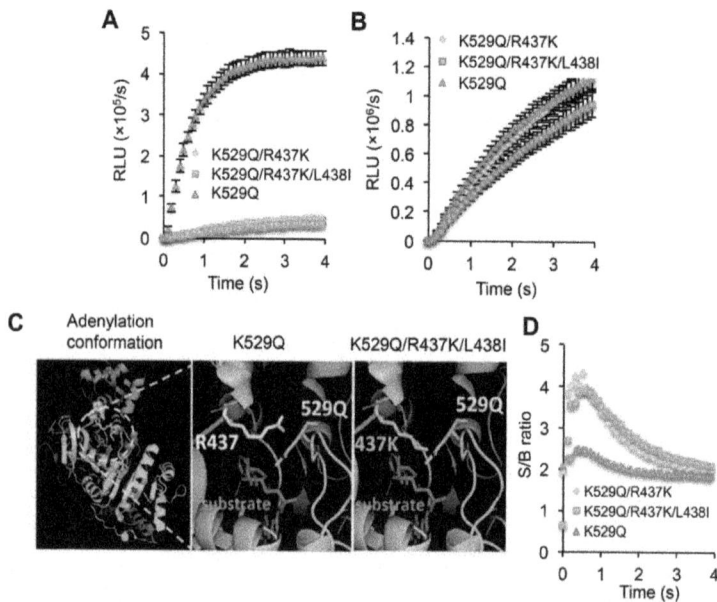

Figure 7. FlimPIA in vitro using the new Acceptor mutated in the hinge region. (A) Overall luminescent activity of the conventional Acceptor and the two new Acceptors. Reactions with LH_2 and ATP (n = 3). (B) Luminescent activity of the Acceptors with LH_2-AMP as a substrate (n = 3). (C) 3D models of the Acceptors at adenylation conformation. The wild-type Fluc (left), the conventional Acceptor (middle), and the mutant M1 (right) are shown. In the conventional Acceptor, the shortest distance between the active site against LH_2 (529Q) and R437 was ~3.8 Å, which was shorter in the mutant (~1.6 Å). (D) FlimPIA with 50 nM each of FKBP/Donor and FRB/the new Acceptor with/ without 50 nM rapamycin (n = 3).

	K_m for LH$_2$	Vmax ($\times 10^4$ RLU/s) for LH$_2$	K_m for ATP	Vmax (10^4 RLU/s) for ATP	K_m for LH$_2$-AMP	Vmax ($\times 10^6$ RLU/s) for LH$_2$-AMP
K529Q	95.0 ± 12.1	3.49 ± 0.20	424 ± 55	2.50 ± 0.11	0.412 ± 0.055	1.04 ± 0.04
K529Q/R437L	115 ± 4.0	5.52 ± 0.08	307 ± 25	3.94 ± 0.11	0.605 ± 0.063	0.737 ± 0.027
K529A/R437K/L438I	62.7 ± 4.1	35.1 ± 0.7	306 ± 25	39.8 ± 1.1	0.710 ± 0.093	1.28 ± 0.06

Table 3. Kinetics properties of Acceptors fused to FRB.

3.5. Optimization of assay conditions

The overall activities of the improved Acceptor (R4437K/K529Q) mentioned in Section 3.4 showed a tenfold decrease, and the oxidative luminescence activities were almost maintained. However, the S/B ratio increased only 1.6-fold. To investigate this discrepancy, the Acceptor was reacted with (1) LH$_2$ + ATP, (2) LH$_2$-AMP, and (3) LH$_2$ + ATP + LH$_2$-AMP (**Figure 8A**). The luminescence intensity in the case of (3) should be equal to the sum of the intensities of (1) and (2). However, the intensity in (3) was remarkably lower than the sum. Therefore, we thought that some competition may exist in the oxidative luminescence steps. It was reported that dehydroluciferyl-AMP (L-AMP), which is converted from LH$_2$-AMP, competes with LH$_2$-AMP in the oxidative luminescence steps, and coenzyme A (CoA) converts L-AMP to dehydroluciferyl-coenzyme A, which is a less potent competitor of LH$_2$-AMP. First, we added

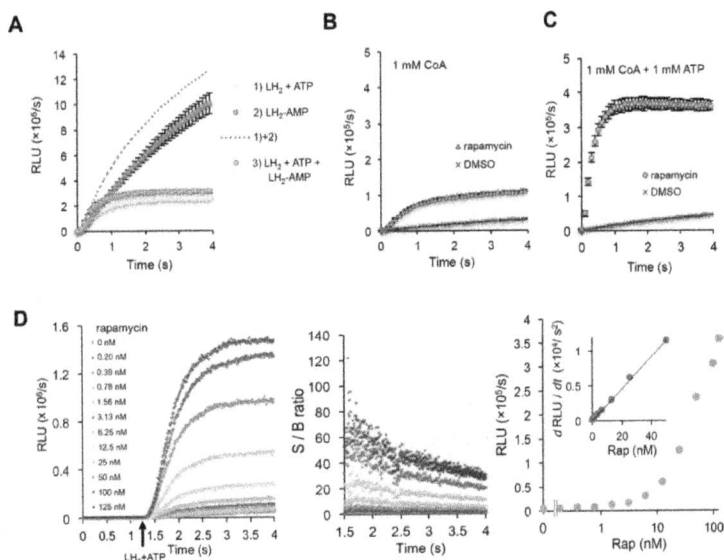

Figure 8. Optimization of assay condition in vitro. (A) An experimental simulation of FlimPIA using the conventional Acceptor. (B) The responses with and without 50 nM rapamycin in the presence of 1 mM CoA and 20 mM ATP. (C) The responses in the presence of 1 mM CoA and 1 mM ATP. (D) The results of tube-based luminometer with rapid mixing of the probes and substrates.

CoA to the mixture of FlimPIA (**Figure 8B**). In the presence of CoA, the maximum S/B ratio reached 8, representing a twofold improvement, when 50 nM of each probe was used.

Next, we optimized the concentration of ATP, as it was designed so that the K_m value of the Acceptor for ATP was lower than that of the wild type to suppress the adenylation activity, but the K_m value of the Donor for ATP was maintained to provide LH_2-AMP. The optimal concentration of ATP was 1 mM, and the maximal S/B ratio reached approximately 40, representing a fivefold improvement, when 50 nM of each probe was used (**Figure 8C**).

Finally, we had optimized the reaction conditions. As the increase of luminescence occurred as soon as substrates were added, a luminometer equipped with a stirrer was used to mix and react the substrates quickly (**Figure 8D**). The luminescence intensity increased quasi-linearly from 0.2 to 0.6 s after the reaction start and then reached a plateau. The maximal S/B ratio reached more than 60 when 100 nM of each probe was used.

Taken together, these improvements achieved a remarkably higher S/B ratio and sensitivity [6].

4. Advantages and disadvantages of FlimPIA

In this section, we describe the advantages and disadvantages of FlimPIA compared to the conventional PPI assay, FRET, and PCA, which are available in cellulo and in vitro.

4.1. FlimPIA in cells

To determine if FlimPIA is applicable in cellulo or in vivo, the FKBP-Donor and FRB-Acceptor were transiently expressed in cultured cells (**Figure 9**) [7]. The response was clearly observed in cells when rapamycin was added, and the luminescence intensity increased depending on the concentration of rapamycin.

However, the maximal S/B ratio was less than 2.5, and the detectable range of the concentration of rapamycin was rather narrow. Although the S/B ratio of FRET is often as low as that of FlimPIA in cells, PCA gives a high S/B ratio both in vitro and in cellulo.

4.2. Stability of probes in vitro

The same Fluc derived from *P. pyralis* was applied to both Fluc-based PCA in vitro and FlimPIA. Then, the thermostability of probes was compared [10]. The probes of Fluc-based PCA (FKBP-C and FRB-N) and the probes of FlimPIA (FKBP-Donor and FRB-Acceptor) were preincubated with or without rapamycin at 37°C (**Figure 10A**). After 30 minutes, half of the luminescence signal was retained in FlimPIA, and on the other hand, the luminescence signal was almost completely diminished in PCA. The rate of the luminescence decay in FlimPIA was approximately one-fourth of the rate of the decay in PCA (**Figure 10B**).

4.3. S/B ratio in vitro

The S/B ratio of FRET is rather low, but on the other hand, PCA shows a high S/B ratio and high sensitivity. The conventional FlimPIA described in Section 3.1 showed that the maximal S/B ratio was 2.5, which is generally lower than the S/B ratio of PCA [7]. However, the S/B ratio dramatically increased by the improvements described in Section 3.2-3.5 and was equal to or higher than the S/B ratio of PCA [6, 9, 10].

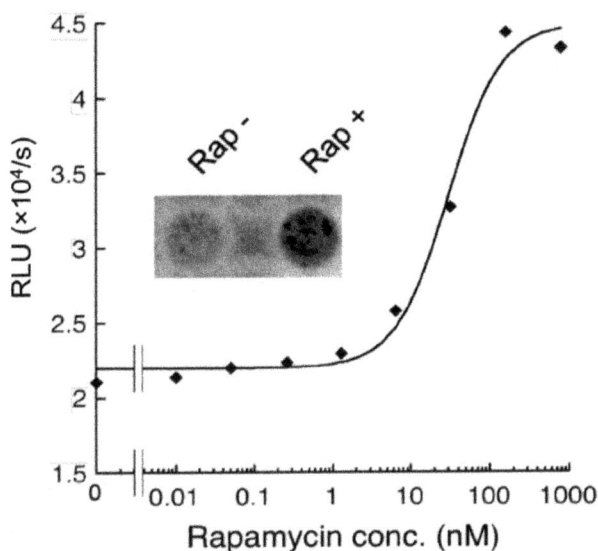

Figure 9. FKBP12-FRB association detected by FlimPIA in cultured cells. ©American Chemical Society.

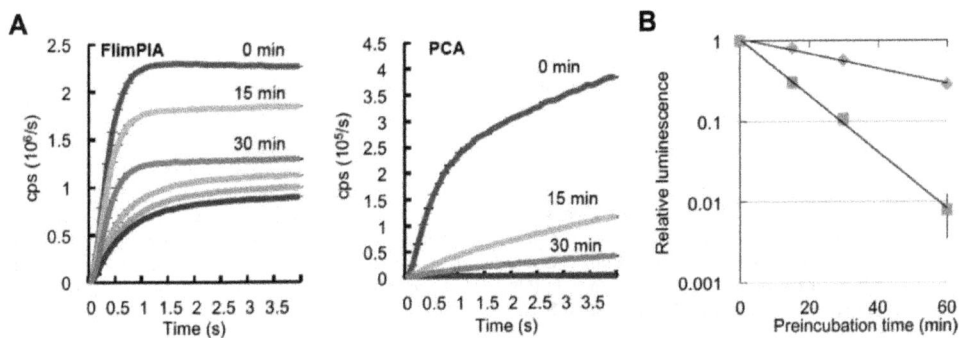

Figure 10. Comparison of thermostability in vitro. (A) Probes (50 nM each) were preincubated at 37°C with or without equimolar rapamycin. The luminescent intensity was measured for 4 s after adding substrates (LH_2 and ATP). Left: FlimPIA, Right: Fluc-based PCA. Red: incubation for 0 min with rapamycin, Orange: 15 min with rapamycin, pink: 30 min with rapamycin, green: 0 min without rapamycin, light blue: 15 min without rapamycin, dark blue: 30 min without rapamycin. (n = 3) (B) Inactivation time course. Relative luminescent intensities at 4 s after reaction start were normalized at the value obtained with 0 min pre-incubation (n = 3). ©American Chemical Society.

4.4. Sensitivity in vitro

The detectable limits of the concentration of rapamycin in Fluc-based PCA, the conventional FlimPIA, and the improved FlimPIA were compared, when 50 nM of each probe (FKBP-C and FRB-N, or FKBP-Donor and FRB-Acceptor) and rapamycin were used. The limits were 250 pM in Fluc-based PCA and 10 pM in FlimPIA using the K529Q/S440L mutant as the Acceptor [4, 9, 10]. The sensitivity of the improved FlimPIA was higher than the sensitivity of Fluc-based PCA.

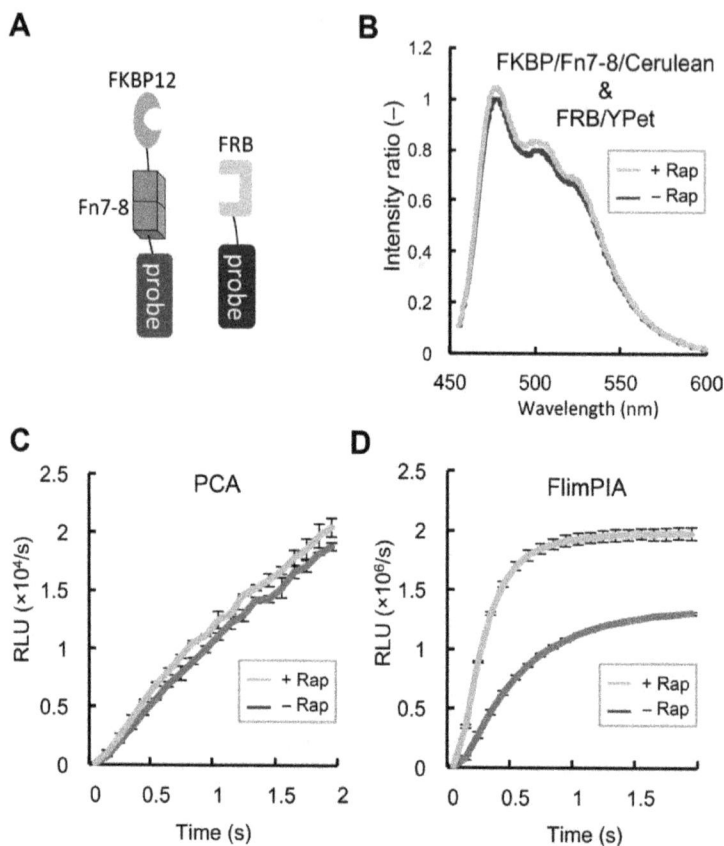

Figure 11. Detectable distance between the probes in vitro. (A) Scheme of the assays. A long (7 nm) Fn7-8 domain is inserted between a binding domain (FKBP12) and a probe. Signals with and without equimolar rapamycin were compared. (B) FRET using 40 nM each of FKBP-Fn7-8-Cerulean and the FRB-YPet as a probe pair. (C) Fluc-PCA using 100 nM each of FKBP-Fn7-8-C and FRB-N (n = 3). (D) FlimPIA using 100 nM each of FKBP-Fn7-8-Donor and FRB-Acceptor (left) (n = 3). ©American Chemical Society.

4.5. Detection limit of dimension of interacting protein in vitro

A fundamental limitation of FRET is that the detectable distance between the two probes is less than several nanometers, because the fluorescent signal is inversely proportional to the sixth power of the distance. A part of fibronectin type III, the seventh and eighth domains (Fn7-8), has a rigid structure with a 7 nm N-C terminal distance [10]. Ohashi et al. reported that a FRET signal using YPet and CyPet could not be observed by inserting Fn7-8 between the two fluorescent proteins [22]. The limit of the detectable distance between the two probes determines the detectable dimensions of the interacting protein.

Therefore, we compared the limit of the detectable distance between the probes in our assay. To examine this, Fn7-8 was inserted between FKBP12 and one of the probes (C-terminal domain for PCA, cerulean for FRET, and Donor for FlimPIA) (**Figure 11**). The large probes were mixed with FRB-N, FRB-YPet, and FRB-Acceptors, respectively. As expected, the FRET signal was very weak when rapamycin was added to the mixture of FKBP12-Fn7-8-Cerulean and FRB-YPet, whereas the signal derived from the mixture of FKBP12-Cerulean and FRB-YPet was clearly observed (not shown). In the case of PCA using FKBP12-Fn7-8-C and FRB-N, the luminescence intensity was not significantly increased by the addition of rapamycin when the concentrations of the probes were moderate (100 nM each), while some response was observed with higher concentrations (750 nM) of each probe (not shown). However, the response of FlimPIA was clearly observed, even when 100 nM each of FKBP-Fn7-8-Donor and FRB-Acceptor was used.

5. Conclusions

We reported the development of Fluc-based PCA using purified probes for the first time. However, the stabilities of the probes were low due to the split forms. The problem might be overcome by using another enzyme with a highly stable structure.

Furthermore, we developed a unique PPI assay, called FlimPIA, wherein the catalytic reaction of Fluc is divided into two half-reactions. FlimPIA has several advantages, especially in vitro. Our next challenge is to improve the response in cellulo.

Acknowledgements

This project was supported partly by SENTAN and SICORP, Japan Science and Technology agency, Japan; by JSPS KAKENHI Grant Numbers JP15H04191, JP17K06920, and JP24040072 from the Japan Society for the Promotion of Science, Japan; by Kikkoman Co.; by Dynamic Alliance for Open Innovation Bridging Human, Environment and Materials from MEXT, Japan; and by the 'Leave a Nest' Microtech-Nichion award.

Conflict of interest

The authors declare that there are no conflicts of interest.

Author details

Yuki Ohmuro-Matsuyama and Hiroshi Ueda*

*Address all correspondence to: ueda@res.titech.ac.jp

Laboratory for Chemistry and Life Science, Institute of Innovative Research, Tokyo Institute of Technology, Yokohama, Japan

References

[1] Venkatesan K, Rual JF, Vazquez A, Stelzl U, Lemmens I, Hirozane-Kishikawa T, Hao T, Zenkner M, Xin X, Goh KI, et al. An empirical framework for binary interactome mapping. Nature Methods. 2009;6:83-90. DOI: 10.1038/nmeth.1280

[2] Zhang QC, Petrey D, Deng L, Qiang L, Shi Y, Thu CA, Bisikirska B, Lefebvre C, Accili D, Hunter T, et al. Structure-based prediction of protein-protein interactions on a genome-wide scale. Nature. 2012;490:556-560. DOI: 10.1038/nature11503

[3] Porter JR, Stains CI, Jester BW, Ghosh I. A general and rapid cell-free approach for the interrogation of protein-protein, protein-DNA, and protein-RNA interactions and their antagonists utilizing split-protein reporters. Journal of the American Chemical Society. 2008;130:6488-6497. DOI: 10.1021/ja7114579

[4] Ohmuro-Matsuyama Y, Chung CI, Ueda H. Demonstration of protein-fragment complementation assay using purified firefly luciferase fragments. BMC Biotechnology. 2013;13:31. DOI: 10.1186/1472-6750-13-31

[5] Dale R, Ohmuro-Matsuyama Y, Ueda H, Kato N. Mathematical model of the firefly luciferase complementation assay reveals a non-linear relationship between the detected luminescence and the affinity of the protein pair being analyzed. PLoS One. 2016;11:e0148256. DOI: 10.1371/journal.pone.0148256

[6] Kurihara M, Ohmuro-Matsuyama Y, Ayabe K, Yamashita T, Yamaji H, Ueda H. Ultra sensitive firefly luciferase-based protein-protein interaction assay (FlimPIA) attained by hinge region engineering and optimized reaction conditions. Biotechnology Journal. 2016;11:91-99. DOI: 10.1002/biot.201500189

[7] Ohmuro-Matsuyama Y, Nakano K, Kimura A, Ayabe K, Ihara M, Wada T, Ueda H. A protein-protein interaction assay based on the functional complementation of mutant firefly luciferases. Analytical Chemistry. 2013;85:7935-7940. DOI: 10.1021/ac4016825

[8] Ohmuro-Matsuyama Y, Ueda H. A protein-protein interaction assay FlimPIA based on the functional complementation of mutant firefly luciferases. Methods in Molecular Biology. 2016;1461:131-142. DOI: 10.1007/978-1-4939-3813-1_10

[9] Ohmuro-Matsuyama Y, Ueda H. Ultrasensitive firefly luminescent intermediate-based protein-protein interaction assay (FlimPIA) based on the functional complementation of mutant firefly luciferases. Methods in Molecular Biology. 2017;1596:119-130. DOI: 10.1007/978-1-4939-6940-1_8

[10] Ohmuro-Matsuyama Y, Hara Y, Ueda H. Improved protein-protein interaction assay FlimPIA by the entrapment of luciferase conformation. Analytical Chemistry. 2014;86:2013-2018. DOI: 10.1021/ac403065v

[11] Chen J, Zheng XF, Brown EJ, Schreiber SL. Identification of an 11-kDa FKBP12-rapamycin-binding domain within the 289-kDa FKBP12-rapamycin-associated protein and characterization of a critical serine residue. Proceedings of the National Academy of Sciences of the United States of America. 1995;92:4947-4951

[12] Chiu MI, Katz H, Berlin V. RAPT1, a mammalian homolog of yeast Tor, interacts with the FKBP12/rapamycin complex. Proceedings of the National Academy of Sciences of the United States of America. 1994;**91**:12574-12578

[13] Paulmurugan R, Gambhir SS. Combinatorial library screening for developing an improved split-firefly luciferase fragment-assisted complementation system for studying protein-protein interactions. Analytical Chemistry. 2007;**79**:2346-2353. DOI: 10.1021/ac062053q

[14] Banaszynski LA, Liu CW, Wandless TJ. Characterization of the FKBP.Rapamycin.FRB ternary complex. Journal of the American Chemical Society. 2005;**127**:4715-4721. DOI: 10.1021/ja043277y

[15] Kussie PH, Gorina S, Marechal V, Elenbaas B, Moreau J, Levine AJ, Pavletich NP. Structure of the MDM2 oncoprotein bound to the p53 tumor suppressor transactivation domain. Science. 1996;**274**:948-953

[16] Branchini BR, Rosenberg JC, Ablamsky DM, Taylor KP, Southworth TL, Linder SJ. Sequential bioluminescence resonance energy transfer-fluorescence resonance energy transfer-based ratiometric protease assays with fusion proteins of firefly luciferase and red fluorescent protein. Analytical Biochemistry. 2011;**414**:239-245. DOI: 10.1016/j.ab.2011.03.031

[17] Sundlov JA, Fontaine DM, Southworth TL, Branchini BR, Gulick AM. Crystal structure of firefly luciferase in a second catalytic conformation supports a domain alternation mechanism. Biochemistry. 2012;**51**:6493-6495. DOI: 10.1021/bi300934s

[18] Ayabe K, Zako T, Ueda H. The role of firefly luciferase C-terminal domain in efficient coupling of adenylation and oxidative steps. FEBS Letters. 2005;**579**:4389-4394. DOI: 10.1016/j.febslet.2005.07.004

[19] Branchini BR, Murtiashaw MH, Magyar RA, Anderson SM. The role of lysine 529, a conserved residue of the acyl-adenylate-forming enzyme superfamily, in firefly luciferase. Biochemistry. 2000;**39**:5433-5440

[20] Branchini BR, Magyar RA, Murtiashaw MH, Anderson SM, Zimmer M. Site-directed mutagenesis of histidine 245 in firefly luciferase: A proposed model of the active site. Biochemistry. 1998;**37**:15311-15319. DOI: 10.1021/bi981150d

[21] Branchini BR, Rosenberg JC, Fontaine DM, Southworth TL, Behney CE, Uzasci L. Bioluminescence is produced from a trapped firefly luciferase conformation predicted by the domain alternation mechanism. Journal of the American Chemical Society. 2011;**133**:11088-11091. DOI: 10.1021/ja2041496

[22] Ohashi T, Galiacy SD, Briscoe G, Erickson HP. An experimental study of GFP-based FRET, with application to intrinsically unstructured proteins. Protein Science. 2007;**16**:1429-1438. DOI: 10.1110/ps.072845607

Protein Interactions and Nanomaterials: A Key Role of the Protein Corona in Nanobiocompatibility

Micaelo Ânia , Rodriguez Emilio , Millan Angel ,
Gongora Rafael and Fuentes Manuel

Additional information is available at the end of the chapter

http://dx.doi.org/10.5772/intechopen.75501

Abstract

The protein corona is still somewhat of a mysterious consequence of the nanoparticles' application in theranostics. In this review, several critical aspects related to the protein corona are described, in particular which influences more specifically its formation, how to evaluate/characterize it, and what interactions to expect when the nanoparticle and the protein corona are inside the cell. Despite these issues, which have been studied in a general way, it has been verified that there's still much information missing when it comes to specific nanoparticles. Here, a few proteins are also highlighted as examples, which have been identified as part of the protein corona; in addition, several factors related to the formation of protein corona are discussed due to their important role in the different adsorbed proteins.

Keywords: protein corona, nanoparticles, nanobiocompatibility, proteomics

1. Introduction

Nanotechnology is becoming everyday a more valuable resource in developing strategies of diagnostics and therapeutics; in fact, a new area is arising which is named nanomedicine [1]. From the use of nanoparticles [2] to nanorobots [3] or nanosensors [4], there is no shortage of ways to apply it to nanomedicine's benefit. Nanoparticles are particularly useful as theranostic agents, as a multifunctional platform which combine both therapeutic and diagnostic applications simultaneously [5]. However, nanoparticles must gather a number of characteristics in order to be considered as good theranostic agents, such as suitable size [6] and shape [7] for cell penetration, biocompatibility, surface charge, efficient targeting [8], and fluorescence, among others [9]. Despite of these advantages and promising applications, there are still many problems associated to the entrance of the nanoparticle in a physiological environment, which may be justified with different intrinsic characteristics of the nanoparticles [10]. In general, there are two different nanoparticle identities, such as "synthetic identity," which refers to their intentional physicochemical properties [11], and a "biological identity," which is related to the physicochemical properties shown by the nanoparticle after its application in a physiological environment and interaction with the presented biomolecules [10]. This "biological identity" is profoundly related to the formation of the protein corona, as it significantly alters the size, shape, and surface charge of the nanoparticle [12, 13]. The protein corona is formed after the entrance of the nanoparticle in a physiological environment, such as the bloodstream and/or peripheral blood, where the presence of thousands of proteins [14] (among other biomolecules) causes their adsorption onto the nanoparticle surface [15], in a corona shape [16]. The formation of this corona is energetically favorable, with a decrease of enthalpy and increase of entropy [15]. It can be divided into two categories: a "hard" corona, and a "soft" corona. The "soft" corona is based on abundant proteins that firstly bind to the nanoparticle through low-affinity bonds, and the "hard" corona is more dense [17], based on sparse proteins that replace the "soft" proteins over time, due to their higher affinity bonds [11], which is known as the Vroman effect [18]. The composition of the corona is directly dependent on the biomolecular composition of the physiological environment that surrounds it [19, 20], the time of exposure [11, 17, 21], and also incubation conditions (such as temperature or mild stirring), among others. Moreover, it is clear that the protein corona is not static and varies in the course of time; in other words, it's dynamic [17, 21]. Eventually, it will reach a state of equilibrium, steady stochastic state, where the association and dissociation rates for each protein occur equally [21], unless it is further incubated in a different biological media or proximal biological fluid, with a formation of a new corona [22]. The great majority of the proteins that form the corona are independent of the size and surface charge of the nanoparticle but are very influenced by the chemical properties of the material that constitutes the nanoparticle [23]. However, there are still sensitive proteins to size and surface charge, whose nature can change and consequently alter the interactions between nanoparticles and cells with the consequent alteration of the biological outcome and biological impacts [23]. For instance, if opsonins bind to the nanoparticle in the protein corona, they will be recognized as a "threat," and consequently are phagocytosed by macrophages [24]. It is crucial to prevent opsonization, "camouflage" the nanoparticle to avoid the phagocytosis, and keep the nanoparticles in circulation, which can be achieved by the application of a polymer coating, such as poly(ethylene glycol) (PEG) [25]. This polymer coating also prevents the formation of the protein corona, which can later compromise the nanoparticle internalization by the cells [25]. It is thus important to study the influence of the protein corona in the internalization of the nanoparticle, as the interactions of the nanoparticles with cells in in vivo studies are much different from the in vitro ones [26–28], which can prove to be an obstacle in the generalized application of this theranostic approach based on nanomaterials.

2. Characterization of protein corona by proteomic strategies

Since the formation of the protein corona has a great impact on the nanoparticle's performance when applied to a biological system, it is important to assess its structure and composition, in order to minimize the adverse effects it may have on the nanoparticle's use. Any alteration in shape, size, electron transfer, or others may come from the binding of the protein corona to the nanoparticle and may be used as parameters of comparison to be tested between nanoparticles, before and after administration to a biological fluid [29]. However, it is necessary to separate the nanoparticle-protein complex from the excess of plasma proteins [30] before assessing the composition of the protein corona. This is frequently made by centrifugation [30, 31], but it can have many adverse effects in the corona, due to the alterations caused by washing steps as well as gradient and volume variations [30, 32, 33]. In order to avoid loss of proteins from the corona, or even tainting the protein corona sample with the proteins in excess from the plasma, centrifugation can be accompanied by other procedures, such as size exclusion chromatography [32] or microfiltration [33, 34]. In the case of magnetic nanoparticles, a one-step centrifugation does not work, as it agglomerates the nanoparticles, making it necessary to perform a magnetic separation [35, 36]. As said by Megido et al. [33], the main methods of evaluation can be held as qualitative or quantitative, being summarized in **Figure 1**.

2.1. Quantitative proteomics

Quantitative proteomics is the collection of techniques that allow the determination of the number of proteins in a sample, which may be its absolute amount or just the relative change

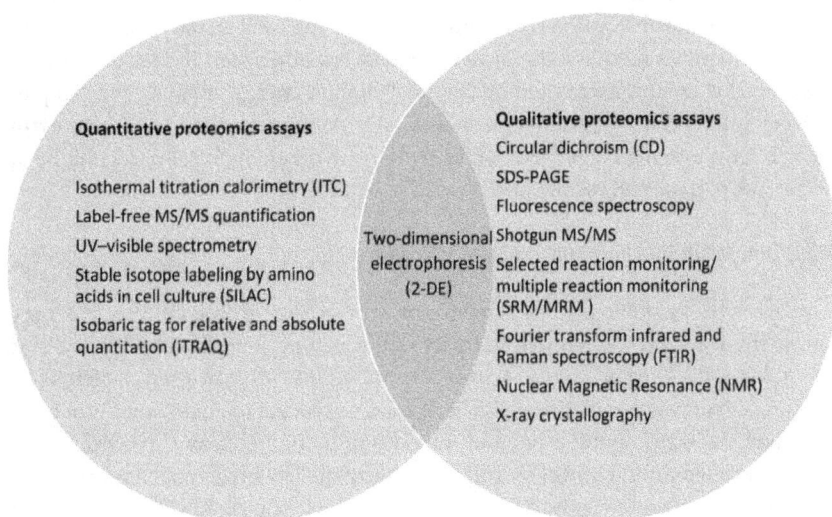

Quantitative proteomics assays

Isothermal titration calorimetry (ITC)
Label-free MS/MS quantification
UV–visible spectrometry
Stable isotope labeling by amino acids in cell culture (SILAC)
Isobaric tag for relative and absolute quantitation (iTRAQ)

Two-dimensional electrophoresis (2-DE)

Qualitative proteomics assays
Circular dichroism (CD)
SDS-PAGE
Fluorescence spectroscopy
Shotgun MS/MS
Selected reaction monitoring/ multiple reaction monitoring (SRM/MRM)
Fourier transform infrared and Raman spectroscopy (FTIR)
Nuclear Magnetic Resonance (NMR)
X-ray crystallography

Figure 1. Summary of methodological approaches useful for the characterization of the protein corona (figure based on Megido et al. [33]).

in amounts between two states [37]. There are many problems associated to the methods used for these assays, such as difficulties in reproducing the results and lack of precision in the measurements [38, 39], but recent technologies have allowed to minimize such issues [40], increasing the depth and coverage [38], which can also be done by using several techniques simultaneously and by defining standards for reproducibility [39]. The most commonly used assays are isothermal titration calorimetry (ITC), UV-visible spectrometry, stable isotope labeling by amino acids in cell culture (SILAC), isobaric tag for relative and absolute quantitation (iTRAQ), and label-free MS/MS quantification. There are also quantitative approaches that make use of two-dimensional electrophoresis (2-DE) [41], but it's mostly used for qualitative proteomics [39], due to the current limitations in performing quantitative assays .

2.1.1. Isothermal titration calorimetry (ITC)

Isothermal titration calorimetry is a method that allows the determination of thermodynamic parameters in a solution (binding affinity, binding stoichiometry, and binding enthalpy change [33]), in particular the ones coming from interactions between biological macromolecules [42]. This process is based on the changes in heat caused by the protein adsorption to the nanoparticle [43]. The main advantage of this method relies on the fact that it allows the characterization while still in the incubation medium [43], which consequently allows a greater optimization of the nanoparticle.

2.1.2. UV-visible spectrometry

UV-visible spectrometry is a process based on the ratio between the passed light measured and the incident light in the UV-visible wavelength [33]. The presence of the protein corona induces changes on the absorption spectrum [17], which makes it an easy, fast, and applicable approach, as it requires no other chemicals or resources other than the protein corona itself [33]. However, it is an unreliable method, as the radiation energy reaching the sample is low [44]; it is very influenced by parameters such as size, temperature, pH [33], and equipment errors, which have a much bigger impact, as there is no other chemical or technique applied to lower the risk of incorrect results [44].

2.1.3. Stable isotope labeling by amino acids in cell culture (SILAC)

SILAC, an acronym to stable isotope labeling by amino acids in cell culture, is a procedure where an essential a.a. has been replaced by its stable isotope counterpart in the cells' growth medium, making this "heavy" amino acid incorporated into all expressed proteins [45]. This causes the growth of two populations of cells: the ones growing in "light" medium containing the natural isotope in the amino acids and the ones growing in "heavy" medium containing stable isotope-labeled amino acids [33, 46]. After complete labeling, equal amounts of labeled and unlabeled cells or protein extracts are mixed in the cell population. The samples are then digested into peptides and then analyzed with mass spectrometry. The quantification of SILAC is thus based on the ratio of introduced isotope-labeled peptides to unlabeled peptides [46]. The many advantages of SILAC are its easy implementation, reasonable quantitative accuracy, and high reproducibility [46].

2.1.4. Isobaric tag for relative and absolute quantitation (iTRAQ)

Isobaric tag for relative and absolute quantitation, also known as iTRAQ, is a widely used method in proteomics for quantification. It is based on mass spectrometry (MS) [47] and is useful in situations where the proteins come from different sources in the same sample [33]. This technique makes use of amine-reactive reagents with different isotopic masses between them [47], labeling the peptides differently and allowing for a clear distinction when analyzing MS scans, as various peptides appear each in a single peak [33]. However, this method has a great disadvantage concerning its cost [48], which makes it impracticable when compared to cheaper alternatives.

2.1.5. Label-free MS/MS quantification

Label-free quantification methods make no use of labeling on the proteins, relying only on the measurement of ion intensity changes in chromatography or on spectrum counting of fragments of peptides in a given protein [46]. This procedure is especially suited for biomarker discovery in large sample sets, as it is not needed labeling in any protein [49]. Labeling also limits the dynamic range, resulting in loss of signal and possible omission of proteins [50]. Therefore, using a label-free quantification approach allows the gathering of reliable information, with great reproducibility [49].

2.2. Qualitative proteomics

Qualitative proteomics refers to the assays designed for identification of proteins in a sample and are often performed not only for identification but also for quantitative purposes, such as the abovementioned 2-DE electrophoresis [39]. Other assessments that allow identification of proteins are circular dichroism (CD), SDS-PAGE, fluorescence spectroscopy, shotgun MS/MS, selected reaction monitoring (SRM)/multiple reaction monitoring (MRM), Fourier transform infrared and Raman spectroscopy, nuclear magnetic resonance (NMR), and X-ray [33].

2.2.1. Circular dichroism

Circular dichroism is an assay based on the determination of the secondary structure, folding and binding properties of proteins [51], using the difference between the absorption of left and right circularly polarized light [52]. This method is based on the optic properties shown by the conformation of the protein, which can be altered when interacting with a nanoparticle. The nanoparticle itself shows no influence in the light, as it is not a chiral compound [33], and it can also be used with small amounts of proteins (20 μg) in a short amount of time [51], making it a viable way of assessment. However, it has some limitations, such as unfeasibility with complex mixtures of proteins [33] and impossibility in obtaining residue-specific information [51].

2.2.2. SDS-PAGE

One of the most used methods in proteomics, electrophoresis, is a procedure that separates proteins in a sample according to their charge. Using a gel of polyacrylamide, a protein solution is applied, and in relation to their charge, proteins will migrate across the gel [53]. The proteins get sorted by molecular weight [53], and staining is needed with an appropriate pigment, for instance, Coomassie blue [54]. The molecular weight is then compared with the one shown by the markers, and a densitometry analysis is performed [33]. This procedure is suitable to characterize the proteins that form the corona, comparing the proteins obtained from the plasma with the ones that are found in the corona. It is thus possible to verify exactly which ones get adsorbed to its surface and, therefore, the ones that have greater affinity to the nanoparticle [43].

2.2.3. Fluorescence spectroscopy

Fluorescence spectroscopy is a method that allows the measurement of the fluorescence of a compound, when excited at a given wavelength [33]. The fluorescence may come from the protein (intrinsic probes), the nanoparticle, or even a fluorophore added to the complex (extrinsic probes) [33, 55], which will be picked up by the amino groups and then detected in the fluorescence spectrometer [56].

2.2.4. Shotgun MS/MS

Shotgun proteomics is a widely used technique in proteomics for identifying proteins [57], with great sensitivity, making it a great influence in the discovery of clinically actionable biomarkers [58]. To perform it, a complex mixture of proteins is separated by sequence-specific proteolysis, forming peptides that will then be separated in smaller fragments to be later analyzed by mass spectrometry [57]. Each peptide will have a mass associated to it, but since peptides may have the same a.a., but in a different arrangement, it is important to assess the sequence, which can be done by the ratios of its mass spectrum, that is, by MS/MS [57].

This method is often disregarded by researchers, due to its early problems in reproducibility and fallible nature [58, 59]. However, the development of bioinformatic tools has allowed a decrease in these problems, making it a viable option for proteomic research [57–59].

2.2.5. Selected reaction monitoring (SRM)/multiple reaction monitoring (MRM)

One of the ways frequently used in replacement of shotgun MS/MS is selected reaction monitoring (SRM) [60], also known as multiple reaction monitoring (MRM). This assay makes use of a triple-quadrupole mass spectrometer [60], where peptides from a previously digested protein go through. In the first analyzer, molecular ions with similar mass to the peptide are selected, followed by fragmentation of the peptide bonds in the second analyzer. Lastly, in the third and final analyzer, the fragmented ions from the peptide pieces are measured, originating a transition signal [61]. This method does not record any full-mass spectrum, increasing its sensitivity and allowing the detection of scarce proteins in complex mixtures [60].

2.2.6. Fourier transform infrared and Raman spectroscopy

FTIR is a procedure that gives information about the surface properties of the nanoparticle and the protein corona, as it allows the detection of its attachment [33]. Standard Raman spectroscopy is not normally used independently, and it's usually meant to complement other methods, such as FTIR [62]. Together, they provide information about protein's secondary structure [63], plus vibrational and rotational parameters.

2.2.7. Nuclear magnetic resonance (NMR)

NMR, which stands for nuclear magnetic resonance, is a form of evaluation of proteins, widely used to describe dendrimers, polymers, and fullerenes derivatives, characterizing structure, purity, and functionality [64] and their possible effects in membrane disruption [65]. Usually, it is used to analyze lipids, as they show high affinity for the nanoparticles [66], after a size exclusion chromatography [33]. It still has disadvantages; for instance, it cannot distinguish the distribution of targeting agent density on a population of nanoparticles [64].

2.2.8. X-ray crystallography

Considered one of the primary sources of structural information about the protein-ligand complex [67], and it is based on the positions and intensities of the reflections as measured in the diffraction pattern of the crystal [67]. This method has big challenges associated to it, as the quantity of radiation needed may be excessive and cause damages to the proteins before a signal is obtained [68], and there are many uncertainties when applying it, such as identity or location of the proteins to be evaluated [67].

3. Nanoparticle's intracellular trafficking

After the nanoparticle's entry in the biological fluid, it is important to ensure its internalization into the cells and intracellular transport, as the formation of the protein corona influences directly the cellular uptake and may also have a significant role on the success of the nanoparticle or lack thereof [69]. Most of mammalian cells internalize the nanoparticles through pinocytosis, although big, specialized cells (such as macrophages) are able to do it by phagocytosis [70, 71], which is the uptake of large particles [70]. Some nanoparticles can also do it by passive penetration of the cellular membrane; however, if the nanoparticle is not small enough, it may deform the membrane [72] by forming holes or thinning it [73], increasing the cytotoxicity [71, 74]. Still, this mechanism is useful in drug delivery, as the nanoparticle travels directly to the cytosol, without making use of endocytic vesicles [74], promoting the reach of the intracellular targets [75]. Hence, it is necessary to take that into account, when designing the nanoparticle, as it may be possible to optimally design the surface of the selected nanoparticle for drug delivery and avoid the membrane's deformation [74]. As for the pinocytosis internalization, which is the cellular uptake of small particles (fluids and solutes), it has four different types of mechanisms [70] (**Figure 2**).

The physical properties of the nanoparticle such as size, net surface charge, and chemical composition determine which endocytosis process is chosen for the internalization [76],

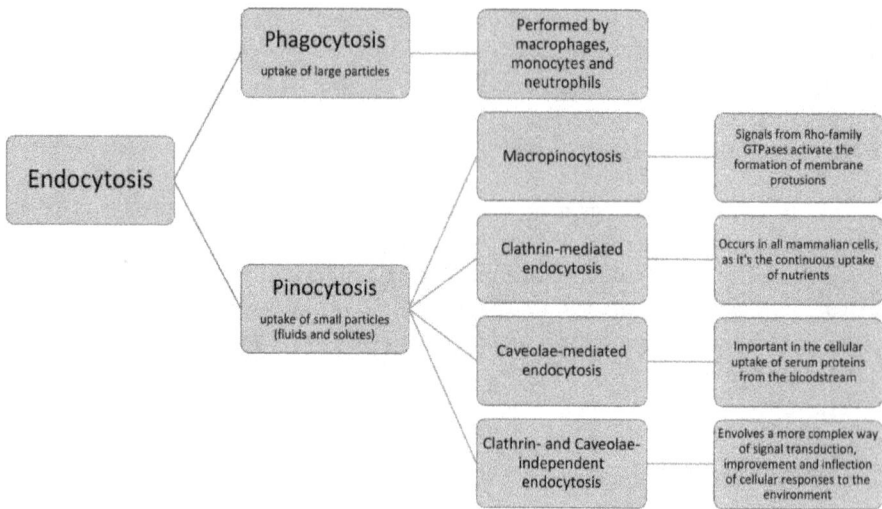

Figure 2. Summary of the different kinds of endocytosis (information based on Conner and Schmid [71]).

although more than one mechanism can be used for the same nanoparticle [77, 78], and the formation of the protein corona may have a great influence in this choice [71]. The proteins adsorbed on the nanoparticle give it its biological identity, and they may present a sequence that is not recognized by the cell as relevant or needed, preventing its endocytosis [79]. It was verified that nanoparticles without a protein corona have higher rates of cellular uptake but can also cause more damage to the cell and alter the cellular metabolism and cell cycle [79, 80]. Nevertheless, if the sequence of proteins in the protein corona is identified as relevant, the endocytosis mechanisms are activated, and the nanoparticle is internalized [79]. When binding to the cellular membrane, the protein corona does not separate itself from the nanoparticle [81], nor does it detach when inside the cell, being internalized as a single complex [82]. After internalization, the nanoparticle's course must be followed by capturing its fluorescence, which can come from the nanoparticle itself or from a fluorescent dye added posteriorly. According to Guarnieri et al. [83], polystyrene nanoparticles follow a fairly diffuse pattern once inside the cell, which suggested no interaction between the nanoparticle and the cytosolic structures, in both situations with and without protein corona. This diffuse pattern can be explained by the nanoparticles being transported within the endocytic vesicles, whose movements are associated to the molecular motors, such as kinesin, myosin, and dynein [83, 84]. Therefore, Guarnieri et al. [83] report that, although the protein corona has some influence in the mechanisms of cellular uptake, it does not show an impact on the intracellular pathways taken by nanoparticles internalized by endocytosis. While leaving the cell, exocytosis mechanisms are activated, and they are dependent on proteins in the medium, because the proteins forming the corona interacted with biological systems inside the cells [76]. The exocytosis is also size, surface coating, and shape dependent, as smaller nanoparticles showed faster exocytosis rates and rod-shaped nanoparticles showed more efficiency when compared to spherical nanoparticles [76, 85]. After performing its function within the cell, the nanoparticle is eventually cleared by the liver and spleen, where they can be kept for a long time, increasing the expected cytotoxicity of the nanoparticle [76].

4. Interaction of nanoparticles with cell interfaces

After internalization, it is important not only to guarantee the achievement of the nanoparticle's function but also to evaluate its effects on cellular organelles [86], as the toxicity cannot be too high, or it will ultimately exclude its use in nanomedicine. The understanding of the nanoparticles' interaction with each cellular organelle is still fairly underdeveloped, as researchers tend to overlook the possible connections between the nanoparticle's composition and the cellular response, focusing considerably more on its uptake [86]. Nonetheless, some studies have already been made to counteract this tendency, in order to give more information and also a better understanding of the nanoparticles' real impact in the cell. According to the experiment performed by Bertoli et al. [78], it is possible to separate the organelles retaining the nanoparticle through magnetism, if the particle is designed to have magnetic properties. Their experiment [78] was based on separating the nanoparticle from the cell after internalization, in order to identify the proteins adsorbed to it, and determining their origin, according to the characteristic proteins from each cellular organelle. The nanoparticles were verified to have the majority of proteins (over 44%) coming from the endocytic pathway, while fewer than 5% came from each of the different organelles studied, such as nucleus, mitochondria, or peroxisomes. However, some proteins can overlap by belonging to more than one organelle [87], acting like a contamination, as they can induce errors in the examination results. Nevertheless, it can be inferred that the majority

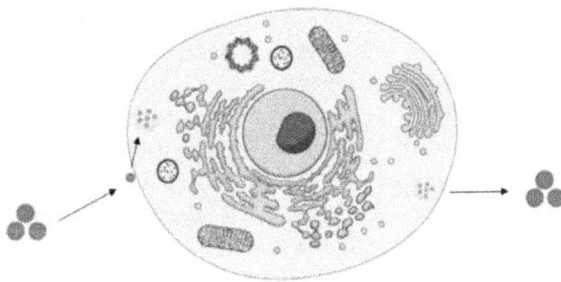

Figure 3. Summarization of the nanoparticle's cellular internalization. The nanoparticles interact with the cell, which can lead to cellular uptake or accumulation of the nanoparticle in the cell membrane. If uptake occurs, the cell engulfs the nanoparticle within endocytic vesicles, which transport the nanoparticle while inside the cell. According to the experiments performed by Bertoli et al. [79], the majority of nanoparticles does not leave the endocytic vesicles to interact directly with other organelles. The nanoparticles are exocytosed afterwards, through lysosomes, to be cleared by the liver or spleen.

of nanoparticles does not leave the endocytic vehicles to associate itself with the different organelles that are present on the cell, which can become a problem, if there is a specific intracellular target for the nanoparticle.

In order to find the internalization's time distribution, Bertoli et al. [78] also performed a time-dependent experiment, evaluating the locations of the nanoparticles after submitting the cells to a short exposure. It was verified that, after only a few minutes, the presence of the nanoparticles was greater in early endocytosis' organelles, with the total absence of lysosomal markers. Still, a more prolonged exposure (24 h) revealed a stronger nanoparticle existence in lysosomes, without any proteins present in early endocytic organelles, indicating their concluding exocytosis from the cells. A similar experiment was previously performed by Shapero et al. [88], where by electron and fluorescence microscopy, the internalization pathway of SiO_2 fluorescent-labeled nanoparticles was characterized. The results obtained also showed a greater number of nanoparticles in the early endocytic organelles after a short exposure, lessening those numbers as time passed and as the nanoparticle's location progressed to lysosomal structures, suggesting its clearance mechanism from the cell. Shapero et al. [88] also verified a nonexistent association between the nanoparticles and the cellular organelles, consolidating the theory that the great majority of nanoparticles does not leave or circulates outside the endocytic vehicles to interact with other organelles (**Figure 3**).

5. A selection of relevant proteins identified in the protein corona

As mentioned before, the constitution of the protein corona is mainly dependent on the composition of the biological medium where it's inserted [19, 20], as different physiological environments have different proteins that compose them. The protein corona is also dependent on the time of exposure [11, 17, 20], chemical properties of the nanoparticle applied [23], and, in some cases, size and surface of the nanoparticle [23]. It is impossible to have a standard protein corona for a given nanoparticle, as each one will have a different composition [15]. The best approach possible then is to characterize the most occurring proteins, in order to predict the behavior of the nanoparticle when inside the physiological system. This issue of research is still emergent, with promising outcomes to better adjust the nanoparticle to its function and environment. An example of protein corona evaluation is the work presented by Mirshafiee et al. [89], who assessed the differences between nanoparticles with different coatings without a protein corona and the same nanoparticles with it adsorbed. Three different types of nanoparticles were used: the bare nanoparticle (with no coating), a nanoparticle with human serum albumin coating (HSA), and a nanoparticle with gamma-globulin coating (GG). The results indicated that there was a different protein corona associated to each of the various coatings, identifying the proteins by LC-MS/MS. For instance, a greater number of lipoproteins and a low quantity of complement factors and immunoglobulins were found in HSA-coated nanoparticles, while the opposite occurred in GG-coated nanoparticles, with higher levels of complement factors and opsonins, especially immunoglobulins, and low levels of lipoproteins. The presence of opsonins made the uptake more difficult, and when comparing with nanoparticles without protein corona, the differences in uptake were very significant, as nanoparticles without it entered the cells more easily.

Another work that explored the composition of the protein corona was provided by Mahmoudi et al. [90], who verified the alteration of the proteins that formed the corona after submitting it to plasmonic heat induction. In this experiment, gold nanorods were used and immersed in fetal bovine serum (FBS) at different concentrations – 10% to mimic in vitro milieu and 100% that mimic in vivo milieu. The protein corona was then evaluated before and after exposure to plasmonic heat induction in both concentrations, and by LC-MS/MS, it was found that, at room temperature, at 10% FBS the most abundant proteins were apolipoprotein A-I precursor and the hemoglobin fetal subunit beta, while the least abundant was the apolipoprotein C-III precursor. At room temperature at 100% FBS, the protein corona was rich in the apolipoprotein A-II precursor and also in hemoglobin fetal subunit beta, while the most scarce was the apolipoprotein C-III precursor as well. After exposure to plasmonic heat induction, at 10% FBS the most abundant proteins became the α-2-HS-glycoprotein precursor and the apolipoprotein A-II precursor, while hemoglobin fetal subunit beta maintaining its numerousness and hemoglobin became the least frequent protein. At 100% FBS after plasmonic heat induction, α-2-HS-glycoprotein precursor also became the most frequent one, followed by hemoglobin fetal subunit beta as well, having a very significant decrease in the quantity of apolipoprotein A-II precursor but still having the apolipoprotein C-III precursor as the least prevalent of all the proteins evaluated. Accordingly with these results [90], it is then possible to conclude that the protein corona's composition is dependent on the medium where the nanoparticle is inserted, which, in this case, also translates to a difference between in vitro and in vivo applications of the nanoparticle, being demonstrated as well its dependency on a physical factor (temperature), which must be taken into account when dealing with hyperthermic nanoparticles as a therapeutic method against tumors.

6. Conclusions

The study of the protein corona is still in a very embryonic stage, with many problems and questions yet to be answered, such as the composition when formed in most nanoparticles, exact description of the uptake and clearance mechanisms, and extensive reports on the consequences of its formation. Some steps have been taken with the purpose of answering these questions, especially in resorting to bioinformatic approaches, allowing an easier and more efficient analysis and sharing of the data obtained. Nevertheless, these are without a doubt interesting research topics, leading the way to improve what is already a very auspicious field in nanomedicine.

Author details

Micaelo Ânia [1], Rodriguez Emilio [2], Millan Angel [3], Gongora Rafael [1,4] and Fuentes Manuel [1,4*]

*Address all correspondence to: mfuentes@usal.es

1 Cancer Research Center (IBMCC/CSIC/USAL/IBSAL), Department of Medicine, General Cytometry Service-Nucleus, University of Salamanca-CSIC, IBSAL, Salamanca, Spain

2 Department of Inorganic Chemistry, Faculty of Pharmacy, University of Salamanca, Salamanca, Spain

3 Department of Condensed Matter Physics and Materials, Research Institute of Aragon, University of Zaragoza, Zaragoza, Spain

4 Proteomics Unit, Cancer Research Center, IBSAL, University of Salamanca, CSIC, Salamanca, Spain

References

[1] Chang EH et al. Nanomedicine: Past, present and future – A global perspective. Biochemical and Biophysical Research Communications. 2015;**468**:511-517

[2] Lee SY, Lin M, Lee A, Park YI. Lanthanide-doped nanoparticles for diagnostic sensing. Nanomaterials. 2017;**7**:1-14

[3] Sim S, Aida T. Swallowing a surgeon: Toward clinical nanorobots. Accounts of Chemical Research. 2017;**50**:492-497

[4] Tang L, Li J. Plasmon-based colorimetric nanosensors for ultrasensitive molecular diagnostics. ACS Sensors. 2017;**2**:857-875

[5] Chen F, Ehlerding EB, Cai W. Theranostic nanoparticles. Journal of Nuclear Medicine. 2014;**55**:1919-1923

[6] Henriksen-Lacey M, Carregal-Romero S, Liz-Marzán LM. Current challenges toward in vitro cellular validation of inorganic nanoparticles. Bioconjugate Chemistry. 2017;**28**:212-221

[7] Champion JA, Katare YK, Mitragotri S. Particle shape: A new design parameter for micro- and nanoscale drug delivery carriers. Journal of Controlled Release. 2007;**121**:3-9

[8] Santi M et al. Rational design of a transferrin-binding peptide sequence tailored to targeted nanoparticle internalization. Bioconjugate Chemistry. 2016;**28**:471-480

[9] Sohaebuddin SK, Thevenot PT, Baker D, Eaton JW, Tang L. Nanomaterial cytotoxicity is composition, size, and cell type dependent. Particle and Fibre Toxicology. 2010;**7**:1-17

[10] Walkey CD, Chan WCW. Understanding and controlling the interaction of nanomaterials with proteins in a physiological environment. Chemical Society Reviews. 2012;**41**: 2780-2799

[11] Lazarovits J, Chen YY, Sykes EA, Chan WCW. Nanoparticle – Blood interactions: The implications on solid tumour targeting. Chemical Communications. 2015;**51**:2756-2767

[12] Dobrovolskaia MA et al. Interaction of colloidal gold nanoparticles with human blood: Effects on particle size and analysis of plasma protein binding profiles. Nanomedicine. 2009;**5**:106-117

[13] Walkey CD, Olsen JB, Guo H, Emili A, Chan WCW. Nanoparticle size and surface chemistry determine serum protein adsorption and macrophage uptake. Journal of the American Chemical Society. 2012;**134**:2139-2147

[14] Anderson NL et al. The human plasma proteome. Molecular & Cellular Proteomics. 2004;**3**:311-326

[15] Ke PC, Lin S, Parak WJ, Davis TP, Caruso F. A decade of the protein corona. ACS Nano. 2017;**11**:11773-11776

[16] Wang B et al. Modulating protein amyloid aggregation with nanomaterials. Environmental Science: Nano. 2017;**4**:1772-1783

[17] Casals E, Pfaller T, Duschl A, Oostingh GJ, Puntes V. Time evolution of the nanoparticle protein corona. ACS Nano. 2010;**4**:3623-3632

[18] Jung S-Y et al. The Vroman effect: A molecular level description of fibrinogen displacement. Journal of the American Chemical Society. 2003;**125**:12782-12786

[19] Cedervall T et al. Understanding the nanoparticle – Protein corona using methods to quantify exchange rates and affinities of proteins for nanoparticles. Proceedings of the National Academy of Sciences. 2007;**104**:2050-2055

[20] Dobrovolskaia MA et al. Protein corona composition does not accurately predict hematocompatibility of colloidal gold nanoparticles. Nanomedicine. 2014;**10**:1453-1463

[21] Dell'Orco D, Lundqvist M, Oslakovic C, Cedervall T, Linse S. Modeling the time evolution of the nanoparticle-protein corona in a body fluid. PLoS One. 2010;**5**:1-8

[22] Lundqvist M et al. The evolution of the protein corona around nanoparticles : A test study. ACS Nano. 2011;**5**:7503-7509

[23] Lundqvist M et al. Nanoparticle size and surface properties determine the protein corona with possible implications for biological impacts. Proceedings of the National Academy of Sciences. 2008;**105**:14265-14270

[24] Owens DE, Peppas NA. Opsonization, biodistribution, and pharmacokinetics of polymeric nanoparticles. International Journal of Pharmaceutics. 2006;**307**:93-102

[25] Dobrovolskaia MA, Aggarwal P, Hall JB, McNeil SE. Preclinical studies to understand nanoparticle interaction with the immune system and its potential effects on nanoparticle biodistribution. Molecular Pharmaceutics. 2008;**5**:487-495

[26] Monopoli MP et al. Physical-chemical aspects of protein corona: Relevance to in vitro and in vivo biological impacts of nanoparticles. Journal of the American Chemical Society. 2011;**133**:2525-2534

[27] Konduru NV et al. Protein corona : Implications for nanoparticle interactions with pulmonary cells. Particle and Fibre Toxicology. 2017;**14**:1-12

[28] Zanganeh S, Spitler R, Erfanzadeh M, Alkilany AM, Mahmoudi M. Protein corona: Opportunities and challenges. The International Journal of Biochemistry & Cell Biology. 2016;**75**:143-147

[29] Li L, Mu Q, Zhang B, Yan B. Analytical strategies for detecting nanoparticle-protein interactions. Analyst. 2010;**135**:1519-1530

[30] Aggarwal P, Hall JB, McLeland CB, Dobrovolskaia MA, McNeil SE. Nanoparticle interaction with plasma proteins as it relates to particle biodistribution, biocompatibility and therapeutic efficacy. Advanced Drug Delivery Reviews. 2009;**61**:428-437

[31] Bewersdorff T, Vonnemann J, Kanik A, Haag R, Haase A. The influence of surface charge on serum protein interaction and cellular uptake: Studies with dendritic polyglycerols and dendritic polyglycerol-coated gold nanoparticles. International Journal of Nanomedicine. 2017;**12**:2001-2019

[32] Cedervall T et al. Detailed identification of plasma proteins adsorbed on copolymer nanoparticles. Angewandte Chemie International Edition. 2007;**46**:5754-5756

[33] Megido L, Díez P, Fuentes M. Nanotechnologies in Preventive and. Regenerative Medicine. 2017:225-244

[34] Bertrand N et al. Mechanistic understanding of in vivo protein corona formation on polymeric nanoparticles and impact on pharmacokinetics. Nature Communications. 2017;**8**:1-8

[35] Bonvin D, Chiappe D, Moniatte M, Hofmann H, Ebersold MM. Methods of protein corona isolation for magnetic nanoparticles. Analyst. 2017;**142**:3805-3815

[36] Sakulkhu U et al. Significance of surface charge and shell material of superparamagnetic Iron oxide nanoparticles (SPIONs) based core/shell nanoparticles on the composition of the protein corona. Biomaterials Science. 2015;**3**:265-278

[37] Ong S-E, Mann M. Mass spectrometry-based proteomics turns quantitative. Nature Chemical Biology. 2005;**1**:252-262

[38] Hackett M. Science, marketing and wishful thinking in quantitative proteomics. Proteomics. 2008;**8**:4618-4623

[39] Choe LH, Lee KH. Quantitative and qualitative measure of intralaboratory two-dimensional protein gel reproducibility and the effects of sample preparation, sample load, and image analysis. Electrophoresis. 2003;**24**:3500-3507

[40] Gallien S, Domon B. Advances in high-resolution quantitative proteomics: Implications for clinical applications. Expert Review of Proteomics. 2015;**12**:489-498

[41] Rabilloud T, Chevallet M, Luche S, Lelong C. Two-dimensional gel electrophoresis in proteomics: Past, present and future. Journal of Proteomics. 2010;**73**:2064-2077

[42] Pierce MM, Raman CS, Nall BT. Isothermal titration calorimetry of protein-protein interactions. Methods. 1999;**19**:213-221

[43] Winzen S et al. Complementary analysis of the hard and soft protein corona: Sample preparation critically effects corona composition. Nanoscale. 2015;**7**:2992-3001

[44] Soovali L, Rõõm E-I, Kutt A, Kaljurand I, Leito I. Uncertainty sources in UV-vis spectrophotometric measurement. Accreditation and Quality Assurance. 2006;**11**:246-255

[45] Waanders LF, Hanke S, Mann M. Top-down quantitation and characterization of SILAC-labeled proteins. Journal of the American Society for Mass Spectrometry. 2007;**18**:2058-2064

[46] Chen X, Wei S, Ji Y, Guo X, Yang F. Quantitative proteomics using SILAC: Principles, applications, and developments. Proteomics. 2015;**15**:3175-3192

[47] Wiese S, Reidegeld KA, Meyer HE, Warscheid B. Protein labeling by iTRAQ: A new tool for quantitative mass spectrometry in proteome research. Proteomics. 2007;**7**:340-350

[48] Zhang H, Wu R. Proteomic profiling of protein corona formed on the surface of nanomaterial. Science China Chemistry. 2015;**58**:780-792

[49] Levin Y, Schwarz E, Wang L, Leweke FM, Bahn S. Label-free LC-MS/MS quantitative proteomics for large-scale biomarker discovery in complex samples. Journal of Separation Science. 2007;**30**:2198-2203

[50] Asara JM, Christofk HR, Freimark LM, Cantley LC. A label-free quantification method by MS/MS TIC compared to SILAC and spectral counting in a proteomics screen. Proteomics. 2008;**8**:994-999

[51] Greenfield NJ. Using circular dichroism spectra to estimate protein secondary structure. Nature Protocols. 2006;**1**:2876-2890

[52] Fleischer CC, Payne CK. Nanoparticle-cell interactions molecular structure of the protein corona and cellular outcome. Accounts of Chemical Research. 2014;**47**:2651-2659

[53] Laemmli UK. Cleavage of structural proteins during the assembly of the head of bacteriophage T4. Nature. 1970;**227**:680-685

[54] Fazekas de St. Groth S, Webster RG, Datyner A. Two new staining procedures for quantitative estimation of proteins on electrophoretic strips. Biochimica et Biophysica Acta. 1962;**71**:377-391

[55] Jameson DM, Croney JC, Moens PDJ. Fluorescence basic concepts, practical aspects, and some anecdotes. Methods in Enzymology. 2003;**360**:1-43

[56] Chen Y, Zhang Y. Fluorescent quantification of amino groups on silica nanoparticle surfaces. Analytical and Bioanalytical Chemistry. 2011;**399**:2503-2509

[57] Marcotte EM. How do shotgun proteomics algorithms identify proteins ? Nature Biotechnology. 2007;**25**:755-757

[58] Theis JD, Dasari S, Vrana JA, Kurtin PJ, Dogan A. Shotgun-proteomics-based clinical testing for diagnosis and classification of amyloidosis. Journal of Mass Spectrometry. 2013;**48**:1067-1077

[59] Searle BC. Scaffold: A bioinformatic tool for validating MS/MS- based proteomic studies. Proteomics. 2010;**10**:1265-1269

[60] Lange V, Picotti P, Domon B, Aebersold R. Selected reaction monitoring for quantitative proteomics: A tutorial. Molecular Systems Biology. 2008;**4**:1-14

[61] Picotti P, Aebersold R. Selected reaction monitoring-based proteomics: Workflows, potential, pitfalls and future directions. Nature Methods. 2015;**9**:555-566

[62] Mandak E, Zhu D, Godany TA, Nystrom L. Fourier transform infrared spectroscopy and Raman spectroscopy as tools for identification of Steryl Ferulates. Journal of Agricultural and Food Chemistry. 2013;**61**:2446-2452

[63] Ridgley DM, Claunch EC, Barone JR. Characterization of large amyloid fibers and tapes with Fourier transform infrared (FT-IR) and Raman spectroscopy. Applied Spectroscopy. 2013;**67**:1417-1426

[64] Patri AK, Dobrovolskaia MA, Stern ST, McNeil SE. Nanotechnology for Cancer Therapy. 2007:105-137

[65] Palermo EF, Lee D-K, Ramamoorthy A, Kuroda K. The role of cationic group structure in membrane binding and disruption by amphiphilic copolymers. The Journal of Physical Chemistry. B. 2012;**115**:366-375

[66] Hellstrand E et al. Complete high-density lipoproteins in nanoparticle corona. The FEBS Journal. 2009;**276**:3372-3381

[67] Davis AM, Teague SJ, Kleywegt GJ. Application and limitations of X-ray crystallographic data in structure-based ligand and drug design. Angewandte Chemie International Edition. 2003;**42**:2718-2736

[68] Chapman HN et al. Femtosecond X-ray protein nanocrystallography. Nature. 2011;**470**:73-78

[69] Kokkinopoulou M, Simon J, Landfester K, Mailander V, Lieberwirth I. Visualization of the protein corona: Towards a biomolecular understanding of nanoparticle-cell-interactions. Nanoscale. 2017;**9**:8858-8870

[70] Conner SD, Schmid SL. Regulated portals of entry into the cell. Nature. 2003;**422**:37-44

[71] Treuel L, Jiang X, Nienhaus GU. New views on cellular uptake and trafficking of manufactured nanoparticles. Journal of The Royal Society Interface. 2013;**10**:1-14

[72] Zhao Y et al. Interaction of mesoporous silica nanoparticles with human red blood cell membranes: Size and surface effects. ACS Nano. 2011;**5**:1366-1375

[73] Leroueil PR et al. Wide varieties of cationic nanoparticles induce defects in supported lipid bilayers. Nano Letters. 2008;**8**:420-424

[74] Wang T, Bai J, Jiang X, Nienhaus GU. Cellular uptake of nanoparticles by membrane penetration : A study combining confocal microscopy with FTIR spectroelectrochemistry. ACS Nano. 2012;**6**:1251-1259

[75] Verma A et al. Surface-structure-regulated cell-membrane penetration by monolayer-protected nanoparticles. Nature Materials. 2008;**7**:588-595

[76] Oh N, Park J-H. Endocytosis and exocytosis of nanoparticles in mammalian cells. International Journal of Nanomedicine. 2014;**9**:51-63

[77] Liu M et al. Real-time visualization of clustering and intracellular transport of gold nanoparticles by correlative imaging. Nature Communications. 2017;**8**:1-10

[78] Bertoli F et al. Magnetic nanoparticles to recover cellular organelles and study the time resolved nanoparticle-cell Interactome throughout uptake. Small. 2014;**10**:3307-3315

[79] Ge C et al. Binding of blood proteins to carbon nanotubes reduces cytotoxicity. Proceedings of the National Academy of Sciences. 2011;**108**:16968-16973

[80] Lesniak A et al. Effects of the presence or absence of a protein corona on silica nanoparticle uptake and impact on cells. ACS Nano. 2012;**6**:5845-5857

[81] Doorley GW, Payne CK. Cellular binding of nanoparticles in the presence of serum proteins. Chemical Communications. 2011;**47**:466-468

[82] Doorley GW, Payne CK. Nanoparticles act as protein carriers during cellular internalization. Chemical Communications. 2012;**48**:2961-2963

[83] Guarnieri D, Guaccio A, Fusco S, Netti PA. Effect of serum proteins on polystyrene nanoparticle uptake and intracellular trafficking in endothelial cells. Journal of Nanoparticle Research. 2011;**13**:4295-4309

[84] Aschenbrenner L, Naccache SN, Hasson T. Uncoated endocytic vesicles require the unconventional myosin, Myo6, for rapid transport through actin barriers. Molecular Biology of the Cell. 2004;**15**:2253-2263

[85] Chithrani BD, Chan WCW. Elucidating the mechanism of cellular uptake and removal of protein-coated gold nanoparticles of different sizes and shapes. Nano Letters. 2007;**7**:1542-1550

[86] Nel AE et al. Understanding biophysicochemical interactions at the nano-bio interface. Nature Publishing Group. 2009;**8**:543-557

[87] Wittrup A et al. Magnetic nanoparticle-based isolation of endocytic vesicles reveals a role of the heat shock protein GRP75 in macromolecular delivery. Proceedings of the National Academy of Sciences. 2010;**107**:13342-13347

[88] Shapero K et al. Time and space resolved uptake study of silica nanoparticles by human cells. Molecular BioSystems. 2011;7:371-378

[89] Mirshafiee V, Kim R, Park S, Mahmoudi M, Kraft ML. Impact of protein pre-coating on the protein corona composition and nanoparticle cellular uptake. Biomaterials. 2016;75:295-304

[90] Mahmoudi M et al. Variation of protein corona composition of gold nanoparticles following plasmonic heating. Nano Letters. 2014;14:6-12

Chapter 4

Protein-Based Detection Methods for Genetically Modified Crops

Kausar Malik, Haleema Sadia and
Muhammad Hamza Basit

Additional information is available at the end of the chapter

http://dx.doi.org/10.5772/intechopen.75520

Abstract

The generation of genetically modified (GM) crops is rapidly expanding each and every year around the world. The well-being and quality assessment of these harvests are vital issues with respect to buyers' interests. This drove the administrative specialists to execute an arrangement of extremely strict strategies for the endorsement to develop and use GMOs and to produce an interest in scientific techniques equipped for identifying GM crops. The GM crops have been added to the effective fuse of various attributes by presenting transgenes, for example, *Bacillus thuringiensis* (Bt) insecticidal qualities, in various crop species. GM crops give critical financial, natural, well-being and social advantages to both small and large agriculturists. The detection strategies incorporate either DNA-based or protein-based measures. Different immunoassays or catalyst connected immunosorbent tests are delicate and more affordable; however, they need experienced technicians. A very simple method, that is, immunochromatographic (ICS) test, is set up in the world, which is modest, compact and simple to utilize. The ICS is a semiquantitative method for indicative screening and semi-measurement of new remote proteins presented through hereditary change of plants. The strip is the easiest method for the assessment of several Bt crop plants for insecticidal quality.

Keywords: immunochromatography, lateral flow strips, detection assay, genetically modified (GM), trait-specific

1. Introduction

The global population is increasing quickly. Experts recommend that food necessities are probably going to rise significantly in the next 20 years. More than 800 million individuals,

IntechOpen

including 33% of the number of inhabitants in sub-Saharan Africa, are undernourished. More than 90% of them are enduring long-term malnutrition and micronutrient insufficiency. Genetic modification of crops can possibly take care of these issues. A hereditarily changed life form (GMO) is a life whose genome has been modified by methods of recombinant DNA technology. This innovation adjusts or embeds at least one quality into a life through genetic modification. GMOs hold extraordinary potential to build trim yield, enhance sustenance quality, decrease input costs and enhance creativity. To date, insect protection and herbicide resilient are the primary business attributes utilized as a part of maize, cotton and soybean [1, 2]. These qualities are giving monetary advantages to the agrochemical business, seed markets and agriculturists because of improved profitability. They additionally conceivably advantage the land because of a lessening in the utilization of chemicals or a move to the utilization of all the more naturally agreeable chemicals.

The development of GM crops is progressing, more qualities are rising and a bigger number of sections of land are being planted with GM crops. The arrival of GM harvests and items in the business sectors worldwide has expanded the administrative need to screen and check the nearness and the measure of GM crops in yields. The worldwide region of GM crops expanded from 1.7 million hectares in 1996 to 81 million hectares in 2004, with an expanding extent developed by creating nations. More than 8 million ranchers are profiting from this innovation [3]. Around 90% of the farmers are small agriculturists from developing nations, who increased their earnings from biotech crops significantly.

The administrative need to screen and confirm the nearness and the measure of GM crops has expanded with the development of the GM crops [4]. Effective monitoring of GM crops must be accomplished with the improvement of proper techniques.

GM crops can be distinguished by identifying the transformed hereditary material at the DNA level, the subsequent protein or phenotype. A few expository techniques, for example, strategies in view of the polymerase chain reaction (PCR) for identifying the incorporated DNA, immunological measures for detecting the subsequent protein or utilizing bioassays to recognize the resultant phenotype have been reported. Western blotting, enzyme-linked immunosorbent assay (ELISA) and parallel stream sticks are common protein-based test techniques [5]. A few other diagnostic advances that can give answers for current specialized issues in the GM test examinations are rising. These techniques incorporate mass spectrometry, chromatography, close infrared spectroscopy, miniaturized scale manufactured gadgets and, specifically, DNA chip innovation (microarrays) and mostly immunoassays. Different immunoassays are being used to determine the genetically modified proteins.

The test on a specimen is, for the most part, a screening test that may distinguish a scope of GMOs. This can be trailed by a particular test to recognize the type of GMO in the sample and additionally intended to measure the quantity of a particular GMO. The lion's share of protein recognition strategies depends on immunoassays for discovery and evaluation of new (outside) proteins presented through the genetic modification of plants. Immunoassay depends on the reaction between an antigen and a counteracting agent. Protein detection strategies for the GMO testing shift from those that are generally modest and simple to perform to more refined measures requiring costly instrumentation. Protein detection strategies can be utilized to recognize GM attributes in GM crops [6]. GMO testing has turned into a vital and essential piece of food production to ensure compliance with labeling regulations, to confirm IP frameworks and secure customers by approving "non-GMO" item publicizing claims [7].

Binding assays are in widely being used in laboratories for the detection and quantification of proteins in samples. For biological samples such as urine, whole blood, plasma, serum and other biological fluids, assays are often performed in hospitals and clinical laboratories. These binding assays can likewise be performed in natural, horticultural, veterinary, mechanical athletic lawful/criminological settings and furthermore, snappy discovery of irresistible sicknesses in serological testing of people and creatures. The principles involved in such assays are well known by those skilled in the art. Many such devices have been described and are available commercially. Immunological binding assay is the sandwich assay. However, in clinical laboratories, the use of solid phase chromatographic binding assay devices has become commonplace for their relative ease of use, economy, and reproducibility. Typically, these chromatographic assay devices are comprised of a porous chromatographic medium which acts as the matrix for the binding assay. The sample is added directly or indirectly to one end of the medium and is chromatographically transported to a detection reagent with which it reacts to form a labeled product, which is then transported to a test zone containing an immobilized capture reagent such as a capture antibody, in which the presence, absence or quantity of an analyte can be determined.

This depends on immunochromatography and sidelong stream measures. It identifies with immunomeasure dipsticks, and especially to those test gadgets used to lead immunological and serological restricting tests. This new strip test is low in cost, quick, monetary, convenient and less laborious. It can be utilized to detect qualitatively or semiquantitatively the presence of protein of interest samples. The development in techniques for utilizing such dipsticks for the detection of GMOs is increasing.

GMOs hold the awesome potential to increase crop yield, enhance nutrient quality, lessen input costs and enhance creativity. To date, insect and herbicide resistance are the principal business attributes utilized in maize, cotton and soybean. The worldwide area of GM crops expanded 47-fold, from 1.7 million hectares in 1996 to 81 million hectares in 2004, with an expanding extent developed by developing nations [2, 3]. GMO testing has turned into an indispensable and essential piece of food production to ensure compliance with labeling regulations. Protein-based methods, for example, ELISA and strip tests are viable for natural items yet rely upon the accessibility of business units and are not appropriate for prepared items because of protein degradation [7]. Parallel stream systems are subjective or semiquantitative [4]. Immunoassays have the ability to be broadly executed on a large scale for the recognition of novel proteins in crude food items. Immunoassay advances are perfect for the subjective and quantitative discovery of many sorts of proteins and pathogens in complex systems [5, 8]. Effective testing of GM products must be accomplished with the improvement of proper strategies for detection. These strategies are for the most part in view of the study of the novel proteins or DNA.

For approval of a scientific strategy, the testing objective must be characterized and execution qualities must be illustrated. Execution qualities incorporate exactness, extraction proficiency, accuracy, reproducibility, affectability, specificity and strength. The utilization of approved strategies is essential to guarantee acknowledgment of results delivered by diagnostic research facilities [9]. The greater part of protein detection techniques depends on immunoassays. Protein detection techniques can possibly recognize the nearness of a particular GM quality and to give the total measurement of the level of transgene expression. Protein identification strategies are exceedingly reasonable for checking particular GM attributes amid treatment of crude items, gave the protein is communicated in the piece of the plant being tried.

Here are the details of immunoassays being used for the detection of genetically modified proteins.

2. Immunoassays

An immunoassay is a biological test that identify and quantify the micro- or macromolecules with the help of antigen or antibody, and the molecule to be detected is called as an analyte. Specific antigens can be stimulated by specific immune responses and as a result of an immune response in the body, antibodies are produced, which are proteins, and they have a sense to find the presence of any foreign antigen in the body. Immunoassays vary in formats. Multiple steps are involved in these assays where reagents are being added and then extra reagents are washed away. Multistep assays are often called heterogeneous immunoassays or separation immunoassays [10]. A few immunoassays can be performed by mixing the samples and reagents and are nonseparation immunoassays or homogeneous immunoassays. The vital component of an immunoassay is an antibody which has a high specificity for the target molecule (antigen), and the area on antigen where antibody attaches is called as an epitope. Standards or calibrators of known concentration are being used to quantify the unknown concentration of analyte. These detections of antigen or antibody take place with the help of labels attached to the antigen or antibody. Many labels are detectable as either they produce a color change in a solution, emit radiations or can be induced to emit light or fluorescence under UV light. The most common used labels for immunoassays are the enzymes.

2.1. History

In the 1950s, the first immunoassay was developed by the Solomon Berson Rosalyn Sussman Yalow. In 1977, Yalow received the Nobel Prize for her work and came in the list of second American women who won this award [11, 12]. In the 1960s [13], the immunoassay became more simple with the discovery of chemically linked enzymes to the antibodies, and later in 1983 [14], Professor Anthony Campbell from Cardiff University introduced acridinium ester in immunoassay that used its own light. This immunoassay helped to quantify a wide range of pathogens, proteins and other proteins in blood samples [14].

3. Classification of immunoassays

1. Competitive homogenous immunoassays

2. Competitive heterogeneous immunoassays

3. One-site noncompetitive immunoassays

4. Two-site noncompetitive immunoassays

3.1. Competitive, homogeneous immunoassays

In competitive homogenous immunoassay, there is a competition between labeled and bound analyte (bound to the antibody) with unlabeled and unbound analyte in the sample. As a competition, the unlabeled and unbound analyte displace the labeled and bound analyte and get them attached in place, while the detached labeled analyte then give fluorescence, and this fluorescence is measured, which is proportional to the amount of unlabeled and initial unbound analyte in the sample.

3.2. Competitive heterogeneous immunoassay

In heterogeneous assay, there is a competition between bound and labeled analyte (bound to the antibody) with unbound and unlabeled analyte, the difference from competitive homogenous assay is that the labeled unbound/displaced analyte is separated by washing, and the remaining labeled and bound analyte is measured.

3.3. One-site noncompetitive immunoassays

In this immunoassay, the amount of unknown analyte in the sample is measured by adding the labeled antibodies. The labeled antibodies get attached with the analyte in the sample, and the extra labeled unbound antibodies are washed away, so, only labeled and bound antibodies are present in the sample, the intensity of fluorescence of these antibodies is measured, which is proportional to the 3.4. amount of unknown analyte in the sample.

3.4. Two-site noncompetitive immunoassays

In this immunoassay, there is an antibody present on a site, and the analyte in a sample is added to the antibody get attached, and then second antibody is added which is attached with the label. If the specific analyte is not present in the solution, the second antibody will not attach. Then, the fluorescence of the labeled antibody is measured, which is directly proportional to the amount of analyte in the sample. It is very important to consider that there are washing steps after every reaction, so extra materials are always washed away. The other thing very important is that what type of labels are attached and how the fluorescence/signal is measured. The details of the labels are given as follows:

3.4.1. Radioactive isotopes

To produce a radioimmunoassay (RIA), radioactive isotopes can be added into the immunoassay reagents and the radiations emitted by bound antigen–antibody complex can be determined by the conventional methods. RIA is considered as the earliest developed immunoassay, and they are not used frequently nowadays because of the hazards of radioactivity [16, 17].

3.4.2. Fluorogenic reporters

Many modern immunoassays are performed by the use of fluorogenic reporters, and the protein microarrays are the best example where these labels are being used [18, 19].

3.4.3. Electrochemiluminescent tags

Electrochemical tags are the labels which emit light as a response of electric current, and the chemiluminescence is detected [20].

3.4.4. DNA reporters

In real-time quantitative PCR, the traditional immunoassays techniques are added and this is called as real-time immunoquantitative PCR (iqPCR). The labels used in this assay are DNA-labeled probes [21, 22].

3.4.5. Enzymes

The most commonly used labels in immunoassays are enzymes, such immunoassays are called as enzyme-linked immunosorbent assays (ELISA) or sometimes enzyme immunoassays (EIAs). Different enzymes are used in such assays, for example, glucose oxidase, horseradish peroxidase (HRM) and alkaline phosphatase. The enzymes are exposed to the reagents which cause them to produce chemiluminescence or light.

4. Label-free immunoassays

There are few immunoassays where labels are not required, for example, in one immunoassay, the antigens are measured by change in resistance in the electrode as the antigen attaches to it. In another method, the binding between unlabeled antibody and antigen is detected by resonance and the technique is called as surface plasmon resonance, and these resonance signals are produced by metal nanoparticle tags which can be measured by a microphone [23, 24].

Different techniques where the immunoassays are being used are as follows:

1. Radioimmunoassay

2. ELISA

3. Memory lymphocyte immunostimulation assay (MELISA)

4. Immunoscreening

5. Cloned enzyme donor immunoassay (CEDIA)

6. Lateral flow test

7. Magnetic immunoassay (MIA)

8. Surround, optical fiber immunoassay (SOFIA)

9. Ultra sensitive antibody detection by agglutination-PCR

10. CD/DVD-based immunoassay.

4.1. Radioimmunoassay

RIA is an extensive method in which radioactive labels are used in a stepwise manner and as it is very specific and sensitive method which require a special equipment. Another such method is called immunoradiometric assay (IRMA) in which radiolabels are used in an immediate manner rather in steps. Radioallergosorbent test (RAST) is used to determine the allergen in case of allergy. It is the cheapest method to perform immunoassay. Although it is the cheapest method to perform immunoassay, it needs special licensing and precautions as radioactive compounds are being used [25–28].

4.1.1. Method

The following steps are required to perform radioimmunoassay:

1. Gamma-radioactive isotopes of iodine, for example, 125-I, attached to the tyrosine are used to label the known amount of antigen

2. A known amount of antibody is mixed with these radiolabelled antigens.

3. Labeled antigen and unlabeled antibody get attached by their binding sites.

4. A sample of serum having the same antigen of unknown amount is added to the mixture.

5. A competition between labeled ("hot") and unlabeled ("cold") antigen is built to attach with antibody binding sites.

6. When the concentration of unknown antigen is increased, it starts to displace the labeled antigen from the antibody binding sites

7. The displaced labeled antigens and bound antigen–antibody complexes got separated, and the radioactivity of displaced radiolabelled antigens is measured by Gamma Counter.

*Radioimmunoassay can be performed as same as the sandwich ELISA method (see sandwich ELISA), the difference is that in ELISA, enzyme is linked with secondary antibody, while in this sandwich radioimmunoassay, radioactive compound is used.

4.2. Enzyme-linked immunosorbent assay (ELISA)

ELISA is a type of immunoassay, and the principle behind its working is the same as that of immunoassay, just it is a wet-lab based assay that uses solid phase enzyme and that is why it is also called as enzyme immunoassay (EIA). ELISA is considered as a quality control test in industries and diagnostic tests in hospitals. ELISA falls under the category of ligand binding assays as it involves the binging of antibody and antigen. When labeled antigens or antibodies get attached to substrates, they make a reaction which causes a change in color, and this color is used as a signal. This substrate to enzyme linkage was developed by Stratis Avrameas and G.B. Pierce. As it is very necessary to wash away the unnecessary or unbound chemicals after each reaction, so that is why the bound antigen-antibody complex should be fixed to the surface of the container with the help of immunosorbent, and this method was developed by Jerker Porath and Wide in 1966. In 1971, a group of different scientists Bauke van Weemen and Anton Schuurs, in the Netherlands, and Eva Engvall and Peter Perlmann, in Sweden, independently published papers describing the methods of ELISA/EIA. Usually, chromogenic reporters and substrates are used, which give observable change in color according to the amount of antigen-antibody complex. In ELISA, a solid phase which is physically immobilized is used to absorb certain components of the liquid phase which has the analyte to be detected. Different reagents and solutions are added, incubated and washed off, and in the end, some optical changes take place which are measured by spectrophotometer at specific wavelength. If the antigen is present in the liquid to be diagnosed/detected, then the labeled antibody is added and vice versa, and then, the substrate is added which reacts with the enzyme of labeled antigen or antibody and then stop solution is added to stop the reaction, and the color change is measured at specific wavelengths by ELISA reader. ELISA can give results in two forms [12, 29–31]:

1. Qualitative: In quantitative ELISA, just positive or negative results can be mentioned. A cutoff value is adjusted by running known positive and negative samples, and the optical density of the solution is measured by spectrophotometer.

2. Quantitative: Quantitative ELISA is used for the quantification of analyte, and the series of standards are used and the unknown amount of analyte is measured.

Different kits are also available in the market for each type of ELISA according to the application and requirement. Mostly, the basic principle and methodology are the same. Procedures and reagents are provided with each specific kit along with the methodology.

Following are the four different types of ELISA and their methodologies:

4.2.1. Direct ELISA

Direct ELISA comprised of the following steps:

- A liquid solution having analyte to be detected is added to the microtiter plate, one sample per well of the plate. The plate has the solid/plastic phase, which absorbs the analyte by charge exchange.

- Bovine serum proteins or casein is added to the wells, which are nonreactive in order to cover that portion of plastic which is not covered by antigen.

- Primary antibody having attached enzyme is added, and it binds with the antigen.

- A substrate is added, which changes the color of the solution by reacting with enzymes.

- The higher the concentration of primary antibody in the solution, the higher the color change will be there and higher will be the analyte in the liquid to be tested.

- The major disadvantage of the direct ELISA is that when antigen is to be measured from serum, antigen mobilization become difficult due to many other proteins present in the serum. Sandwich or indirect ELISA becomes more suitable in that case (**Figure 1**).

4.2.2. Sandwich ELISA

Sandwich ELISA is a type of immunosorbent assay in which one antigen is sandwiched between the two antibodies or one antibody is sandwiched between two antigens for more specific reactions. The procedure of the ELISA is given below [32, 33]:

1. A known amount of antibody is bound to a fixed surface.

2. Nonspecific sites on solid surface are blocked by bovine serum albumin, casein or any other such neutral solution.

3. The sample containing antigen is applied to the plate and which is captured by antibody.

4. The unbound antigens are washed away by washing solution

5. The secondary antibody is added which is also labeled with enzymes.

6. The unbound antibodies are washed away.

7. The sandwich is formed having two antibodies and one antigen inside.

8. A substrate is added, and the enzyme reacts with the substrate and gives a color which is proportional to the amount of antigen.

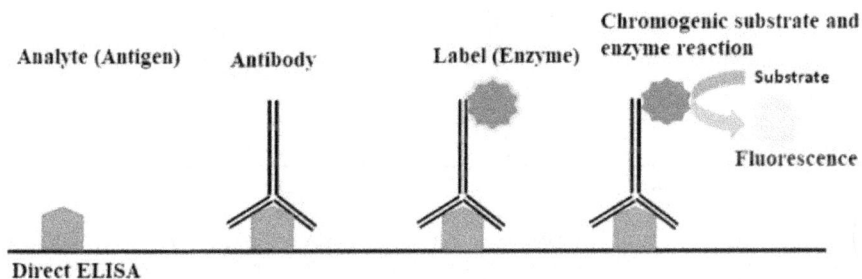

Figure 1. Direct ELISA. This figure shows the direct ELISA in which the analyte (antigen) to be determined is attached with the labelled antibody and then chromogenic substrate is added which reacts with the enzyme and gives fluorescence.

9. The absorbance or fluorescence or electrochemical signal (e.g., current) of samples is measured to determine the presence of antigen and to quantify it (**Figure 2**).

4.2.3. Indirect ELISA

Indirect ELISA is the same as that of the direct ELISA, only primary antibody has been unlabeled, which is very specific to the antigen and the labeled secondary antibody is added,
which is labeled withe enzyme or any other label (**Figure 3**).

4.2.4. Competitive ELISA

There is a competition of analyte in this ELISA. The procedure is as follows:

1. First sample having antigen is incubated in the presence of antibody.

2. This antigen–antibody complex is then added into the antigen-coated well.

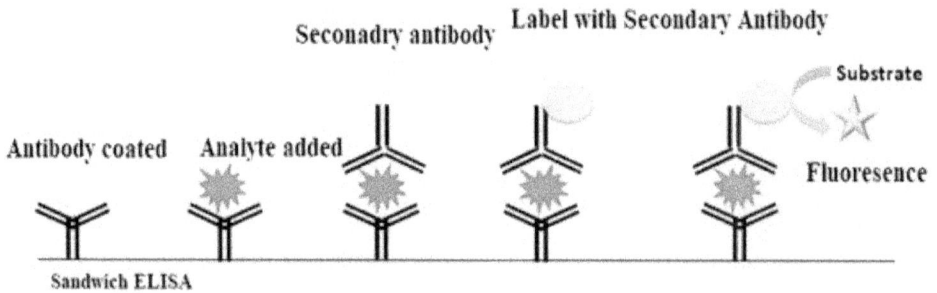

Figure 2. Sandwich ELISA. This figure shows the antigen is the analyte which is sandwiched between two antibodies,
the antibody can be sandwiched between two antigens in the same way.

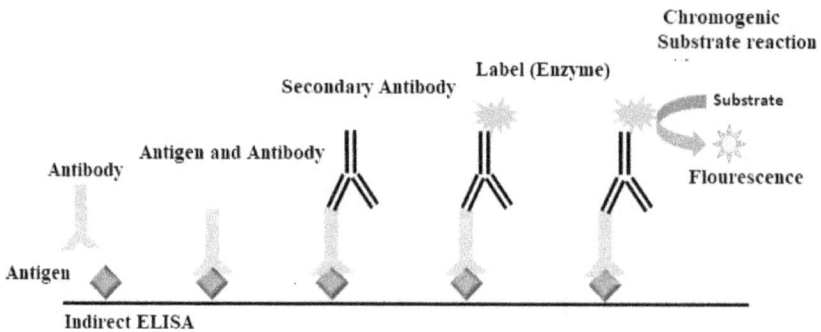

Figure 3. Indirect ELISA. This figure shows the antigen is the analyte to be detected, the primary antibody specific to
the antigen is added, secondary antibody is added then, which is labelled with the enzyme and after that chromogenic
substrate is added, which reacts with the enzyme and give fluorescence.

3. The plate is washed to remove all the unbound antibodies, and the antigen-antibody (Ag-Ab) complex have a competition with labeled antigen as there are less unbound antibodies and coated antigen needs to attach with antibodies which will be taken from bound Ag-Ab complex.

4. Then, the secondary antibody is added which is attached with the enzyme.

5. A substrate is added, and as a reaction of the enzyme and substrate, color is produced.

6. To prevent the eventual saturation of the signal, the stop solution is added to stop the reaction.

In some kits, enzyme-linked antigens are used instead of antibodies, and the remaining competition mechanism is same as described earlier; therefore, there will be a competition of antigens instead of antibodies.

4.2.5. Applications of ELISA

To determine the immune response in the body enzyme-linked immunosorbent assay (ELISA) is being used, which have different methods. These methods are being extensively used to determine analyte in the biological samples of whole blood, serum, urine and other biological fluids. These assays have wide applications in agriculture, industrial, environmental, athletic and legal/forensic fields [34–38].

This assay can be used to determine:

1. The antigen present in oncology samples. The elevated levels of carcinoembryonic antigen (CEA) and prostate-specific antigen (PSA) can be used for early diagnosis of tumorigenic processes.

2. Other disorders and diseases can be diagnosed by immunoassays, for example, antigenic determinants of infectious disease organisms, including fungi, viruses and bacteria and yeast. *Helicobacter pylori*, Mycobacterium tuberculosis, malaria, West Nile Virus, HIV, human papilloma virus (HPV), human chorionic gonadotropin (HCG) for pregnancy, hormones determination, gastrointestinal disorders determination, inherited metabolic diseases determination by enzymes, hexosaminidase as a marker of Tay-Sachs disease and histidase as a marker of histidinemia, tissue damages determination by tissue specific antigen in circulation by determination of creatinine kinase for muscle damage cardiac troponine for myocardial infarction.

3. Suitable identification of the foreign protein in genetically modified organisms (GMOs) and antibodies.

4. To test the athletes' blood sample for recombinant growth hormone, immunoassays are widely used in sports anti-doping laboratories (rGH rhGH, GH, hGH).

4.3. Memory lymphocyte immunostimulation assay (MELISA)

Type-IV hypersensitivity to chemicals, metals and environmental toxins such as molds can be determined by an immunoassay called as a memory lymphocyte immunostimulation assay (MELISA). The test determines the harmful substance in the blood, which is causing allergic reactions, but it will not measure the amounts of toxic substances. Two research articles showed MELISA had many false positive results, while one subsequent study showed that it is very reliable, specific and sensitive method to detect the metals in metal allergic patients [39–43].

4.4. Immunoscreening

It is a method to determine the proteins produced by genes inserted into expression vectors. For this, antiserum should be available and the secondary antibody should also be labeled with radioactive compounds or enzymes [44].

4.5. Cloned enzyme donor immunoassay

In this type of Immunoassay, in which two types of enzymes are being used which can be active only when they combine together [45]. The one enzyme is conjugated with the same type of specific analyte to be determined and this enzyme complex is called as analyte-enzyme-fragment conjugate. The other enzyme attaches to the specific antibody. The analyte-enzyme-fragment conjugate is unable to assemble with the other enzyme, if it is attached to the antibody. For this purpose, the antibody should be displaced from the enzyme.

Therefore, when the analyte to be determined present in the serum is mixed with the analyte-enzyme-fragment conjugate and antibody-enzyme. There is a competition between the analyte in the serum and the analyte-enzyme-fragment conjugate. If the concentration of analyte is high in the serum, then, it will attach with the antibody-enzyme, and the enzyme will be free to attach with the analyte-enzyme-fragment conjugate to give enzyme activity with the substrate. It means the higher the concentration of analyte in the serum, the higher will be enzyme activity and vice versa.

4.6. Lateral flow immunochromatographic assays

Simple devices (Strips) are being used to detect the analyte of interest in the sample without the need of any equipment. A widely used such tests are home pregnancy test, HCV, HBV diagnostic test, and so on. Immunoassays have the ability to be broadly executed on a business scale for the recognition of proteins in food [10, 46–49]. The test is used to detect Bt-GM crops for the expression of insecticidal crystal protein (ICP) of *Bacillus thuringiensis*. One-step lateral flow tests, which are also called immunochromatographic strips (ICS) or dipstick tests, have been a popular platform for qualitative rapid visual tests, which use colloidal gold conjugate to generate signals.

In previous methodologies, QuickStix lateral flow test devices employ the same immunoassay principles as the plate format, but coat the antibodies and other reagents on a nitrocellulose membrane rather than on the inside of test wells or tubes. Nitrocellulose (NC) membranes have been the first choice of device manufacturers for over 20 years. A test strip assay device, in which a mobile conjugate labeled with colloidal labels such as gold, can be deposited on a chromatographic medium, and after reaction with an analyte, thus transported with the solvent to a test zone. The labeled mobilizable detection reagent reacts with an analyte, and the resulting product migrates with the liquid sample as the sample progresses to the test zone. During manufacturing, after the unlabeled binding agent is added to and immobilized in the test zone, the remainder of the test strip material is treated with blocking agents, in order to block any remaining binding sites. The zone where the mobilizable labeled reagent is located is often referred to as the "labeling zone," but can be referred to as the "reversible immobilization zone" or "mobilization zone" while the analyte is reacting with the mobilized labeled reagent, the liquid sample and mobilized labeled reagent migrates further within the porous carrier to the detection zone, where reagent that binds the same analyte is fixed or immobilized, usually in the form of a line. The important aspects of antibody pairs include steric separation of epitopes, an adequate titer of stocks, high affinity, high specificity, high avidity and purity.

The benefits of immunochromatographic tests include user-friendly format, very short time to get a test result, long-term stability over a wide range of climates and relatively inexpensive to make. These features make strip tests ideal for applications such as home testing, rapid point of care testing and testing in the field for various environmental and agricultural analytes. It is limited to diagnostic screening applications only. Furthermore, the achievable sensitivity is a factor of about 10–100 poorer than an instrumented laboratory immunoassay, restricting the technology's utility to relatively high abundance analytes only. Some of the more common lateral flow tests currently on the market are tested for pregnancy, strep throat [50], Chlamydia and human brucellosis [51]. Lateral flow assays have been used extensively as diagnostic tools for monitoring of toxins (**Figure 4**).

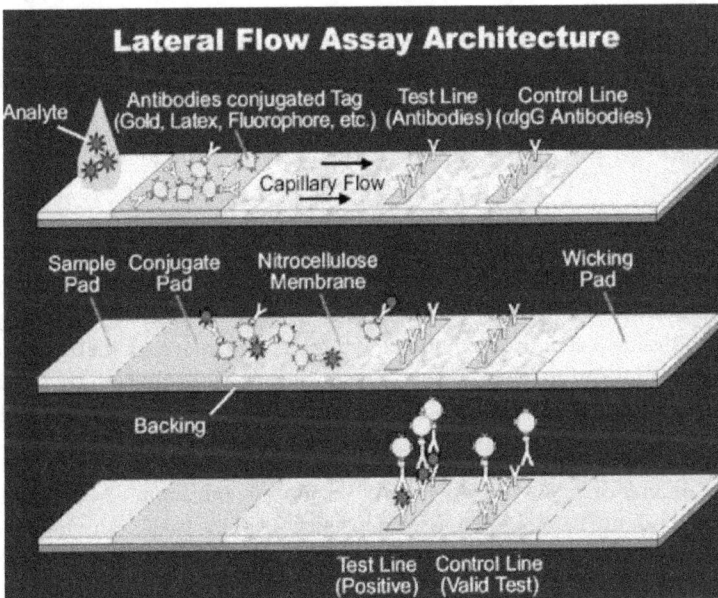

Figure 4. Lateral flow method.

4.7. Magnetic immunoassay (MIA)

The magnetic nanoparticles were discovered by Frenchman Louis Néel, and he got the first Nobel Prize in Physics in 1970. The scientists described the superparamagnetic quality of these magnetic nanoparticles in the magnetic field. These component magnetic nanoparticles are in the range of 5–50 nm while the magnetic beads may be in the range of 35 nm–4.5 μm. A novel type of diagnostic immunoassay was developed by using these magnetic beads as labels. The presence of magnetic labels is measured by the magnetic reader, that is, magnetometer. Therefore, the signals measured by the instrument are directly proportional to the analyte in the serum (toxin, cardiac marker, virus, bacteria). The superparamagnetic quality of these beads has already been in practice in magnetic resonance imaging (MRI) [52].

4.8. Surround, optical fiber immunoassay (SOFIA)

A billion times more sensitive and dynamic technique than conventional diagnostic methods is "surround, optical fiber immunoassay (SOFIA)" for in vitro diagnostics, in which surround optic fiber assembly is used to capture the fluorescence from the sample. SOFIA's sensitivity is up to attograms level, that is, (10^{-18} g). SOFIA has a power to differentiate the analyte over 10 orders of magnitude. This technique is used for *ante mortem* screening test for, Scrapie, BSE, vCJD, CWD, Alzheimer's, Parkinson's disease and transmissible spongiform encephalopathies [53].

4.9. Ultra sensitive antibody detection by agglutination-PCR (ADAP)

With this technique, the antibodies in the ultrasensitive solution are detected by synthetic antigen-DNA conjugates, which enable the ligation of strands of DNA, and quantification is done by qPCR. ADAP can detect zepto- to attomoles of antibodies with dynamic range of 5–6 orders of magnitude in 2 μL of the sample. Agglutination-PCR gives 1000-fold increased sensitive results in the determination of the anti-thyroglobulin autoantibodies from human patient plasma. The ADAP is very sensitive, and very cheap equipment such as Slip Chip is being used, and there is no need to use hazardous radioactive compounds [54].

4.10. CD/DVD-based immunoassay

Storage and retrieval of information can be performed on the metal reflective layer and the polycarbonate surface of CD/DVD. The metal surface of the CD is made of pure gold sometimes, and it shows perfect optical activities and this metal can perform the activity of the substrate and compounds can attach to it and as a result, it can change the refractive and reflective properties of the disk, and the signals produced can tell the amount of analyte in the sample.

In addition to the abovementioned immunoassays, there are many other ELISA-based immunoassays, the difference is that ELISA is used to determine the analyte in the liquid solution while these methods are being used to determine analyte in the tissue samples after performing a series of steps, provided with easy time of assay, for example, Western blot, immunohistochemistry, dot blot, immunocytometry, immunostaining. It is very important to know that in immunoassays, there is an importance of antibodies of immunoglobulin, but scientists are working hard to make this procedure more cheaper.

5. Aptamers

Aptamers are single-stranded oligonucleotides of DNA or RNA molecules, and have property to bind with high affinity and specificity to their target due to their strong interactions and nanosize, respectively. This property of aptamers can be used for a number of applications in biomedical research, their high efficiency of molecular recognition makes them effective biosensors and therefore, they can be used to develop assays against different targets [55]. Different aptamers can be synthesized for a specific target through a process called systematic evolution of ligands by exponential enrichment (SELEX). Biosensing property of aptamers offers fast and easy detection of target molecules. This property can be used for diagnosis and other biomedical applications, which will help to fight against a number of diseases, including AIDS, cancer, Alzheimer's, viral and bacterial infections. A number of aptamers can be identified against various targets, including nucleotides, proteins, lipids, signaling molecules and even whole cells and microorganisms. Recent advances in research have proven that RNA aptamers have high therapeutic and diagnostic value. It can also be used for therapeutic delivery of oligos. All these attributes of aptamers make them pivotal tools of the emerging bionanotechnology and biosensors. Some research groups are working on aptamer technology and using them as aptasensors but it requires more attention to boost our research for diagnosis and fight against different diseases. Aptamers are easy to synthesize and more stable as compared to antibodies; therefore, they can be helpful in our future advances in therapeutics and diagnosis [56].

Author details

Kausar Malik[1]*, Haleema Sadia[2] and Muhammad Hamza Basit[1]

*Address all correspondence to: kausarbasit786@yahoo.com

1 National Centre of Excellence in Molecular Biology, University of the Punjab, Lahore, Pakistan

2 Department of Biotechnology and Informatics, Balochistan University of Information Technology, Engineering and Mangement Sciences, Lahore, Pakistan

References

[1] Dunwell JM. Novel food products from genetically modified crop plants: methods and future prospects. International Journal of Food Science and Technology. 1998;**33**(3): 205-213

[2] James C. Preview: Global status of commercialized transgenic crops: ISAAA briefs No. 30. Ithaca, NY: ISAAA; 2003

[3] Matsuoka T, Kuribara H, Akiyama H, Miura H, Goda Y, James C. Preview: Global Status of Commercialized Biotech/GM Crops; 2004

[4] Tripathi L. Review: Techniques for detecting genetically modified crops and products. African Journal of Biotechnology. 2005;**4**(13):1472-1479

[5] Brett GM, Chambers SJ, Huang L, Morgan MRA. Design and development of immunoassays for detection of proteins. Food Control. 1999;**10**(6):401-406

[6] Stave JW, Magin K, Schimmel H, Lawruk TS, Wehling P, Bridges A. AACC collaborative study of a protein method for detection of genetically modified corn. Cereal Foods World. 2000;**45**:497-501

[7] Spiegelhalter F, Lauter F-R, Russell JM. Detection of genetically modified food products in a commercial laboratory. Journal of Food Science. 2001;**66**(5):634-640

[8] Ahmed FE. Detection of genetically modified organisms in foods. Trends in Biotechnology. 2002;**20**:215-223

[9] Lipp M, Anklam E, Stave JW. Validation of an immunoassay for detection and quantification of genetically modified soybean in food and food fractions using reference materials: Interlab Study. Journal of AOAC International. 2000;**83**:919-927

[10] Yetisen AK. Paper-based microfluidic point-of-care diagnostic devices. Lab on a Chip. 2013;**13**(12):2210-2251. DOI: 10.1039/C3LC50169H

[11] Rall JE, Berson SA. In "Biographical Memoirs". National Academy of Sciences. 1990;**59**:54-71. ISBN: 0-309-04198-8. Fulltext

[12] Yalow RS, Berson SA. Immunoassay of endogenous plasma insulin in man. The Journal of Clinical Investigation. 1960;**39**:1157-1175. DOI: 10.1172/JCI104130. PMC 441860. PMID: 13846364

[13] Lequin R. Enzyme immunoassay (EIA)/enzyme-linked immunosorbent assay (ELISA). Clinical Chemistry. 2005;**51**(12):2415-2418. DOI: 10.1373/clinchem.2005.051532. PMID: 16179424

[14] Prof Anthony Campbell - MA PhD. Cardiff University. Retrieved 29 December 2012

[15] NPS Focus. Rainbow Makers. Royal Society of Chemistry (RSC); 2003. Retrieved 29 December 2012

[16] Landers SJ. ELISA test marks 35 years of answering medical questions. American Medical News. 3 April 2006. Retrieved 9 December 2012

[17] Yalow RS. America.gov. April 27, 2008. Retrieved June 26, 2010

[18] Rajkovic E-M. Immunoquantitative real-time PCR for detection and quantification of Staphylococcus aureus enterotoxin B in foods". Applied and Environmental Microbiology. 2006;**72**(10):6593-6599. DOI: 10.1128/AEM.03068-05. PMC 1610299. PMID: 17021210

[19] Gofflot El. Immuno-quantitative polymerase chain reaction for detection and quantitation of prion protein. Journal of Immunoassay and Immunochemistry. 2004;**25**(3): 241-258. DOI: 10.1081/ias-200028044. PMID: 15461386

[20] Luminex xMAP Technology. Millipore Corporation. Retrieved 13 December 2012

[21] Chatterjee S. Protein microarray on-demand: a novel protein microarray system. PLoS ONE. 2008;3:e3265. DOI: 10.1371/journal.pone.0003265. PMC 2533396. PMID: 18813342

[22] Forster RJ, Bertoncello P, Keyes TE. Electrogenerated chemiluminescence. Annual Review of Analytical Chemistry. 2009;2:359-385. DOI: 10.1146/annurev-anchem-060908-155305. PMID: 20636067

[23] González-Díaz JB et al. Plasmonic Au/Co/Au nanosandwiches with enhanced magneto-optical activity. Small. 2008;4(2):202-205. DOI: 10.1002/smll.200700594. PMID: 18196506

[24] Tsekenis G. Label-free immunosensor assay for myelin basic protein based upon an ac impedance protocol. Analytical Chemistry. 2008;80(6):2058-2062. DOI: 10.1021/ac702070e

[25] Vare EA, Ptacek G. Patently Female: From AZT to TV Dinners: Stories of Women Inventors and Their Breakthrough Ideas. New York: Wiley; 2002. p. 99. ISBN: 0471023345

[26] Radioimmunoassay at the US National Library of Medicine Medical Subject Headings (MeSH)

[27] http://users.rcn.com/jkimball.ma.ultranet/BiologyPages/R/Radioimmunoassay.html

[28] http://www.biomnis.com/en

[29] Wide L, Porath J. Radioimmunoassay of proteins with the use of Sephadex-coupled antibodies. Biochimica et Biophysica Acta. 1966;130(1):257-260. DOI: 10.1016/0304-4165(66)90032-8

[30] Engvall E, Perlmann P. Enzyme-linked immunosorbent assay (ELISA) quantitative assay of immunoglobulin G. Immunochemistry. 1971;8(9):871-874. DOI: 10.1016/0019-2791(71)90454-X. PMID: 5135623

[31] Van Weemen BK, Schuurs AHWM. Immunoassay using antigen—Enzyme conjugates. FEBS Letters. 1971;15(3):232-236. DOI: 10.1016/0014-5793(71)80319-8

[32] Schmidt SD, Mazzella MJ, Nixon RA, Mathews PM. Aβ measurement by enzyme-linked immunosorbent assay. Methods in Molecular Biology. 2012;849:507-527. DOI: 10.1007/978-1-61779-551-0_34. PMID: 22528112

[33] Kragstrup TW, Vorup-Jensen T, Deleuran B, Hvid M. A simple set of validation steps identifies and removes false results in a sandwich enzyme-linked immunosorbent assay caused by anti-animal IgG antibodies in plasma from arthritis patients. 2013

[34] MedlinePlus Encyclopedia ELISA/Western blot tests for HIV

[35] Food Allergen Partnership" (Press release). FDA. January 2001. Retrieved August 20, 2015

[36] Sblattero D, Berti I, Trevisiol C, Marzari R, Tommasini A, Bradbury A, Fasano A, Ventura A, Not T. Human recombinant tissue transglutaminase ELISA: An innovative diagnostic assay for celiac disease. The American Journal of Gastroenterology. 2000;95(5): 1253-1257. DOI: 10.1111/j.1572-0241.2000.02018.x. PMID: 10811336

[37] Porcelli B, Ferretti F, Vindigni C, Terzuoli L. Assessment of a test for the screening and diagnosis of celiac disease. Journal of Clinical Laboratory Analysis. 2016 Jan;**30**(1):65-70. DOI: 10.1002/jcla.21816

[38] Griffin JFT, Spittle E, Rodgers CR, Liggett S, Cooper M, Bakker D, Bannantine JP. Immunoglobulin G1 enzyme-linked immunosorbent assay for diagnosis of Johne's disease in red deer (*Cervus elaphus*). Clinical and Vaccine Immunology. 2005;**12**(12):1401-1409. DOI: 10.1128/CDLI.12.12.1401-1409.2005. PMC 1317074. PMID: 16339063

[39] Cederbrant K, Gunnarsson LG, Hultman P, Norda R, Tibbling-Grahn L. In vitro lymphoproliferative assays with $HgCl_2$ cannot identify patients with systemic symptoms attributed to dental amalgam. Journal of Dental Research. Aug 1999;**78**(8):1450-1458. DOI: 10.1177/00220345990780081101. PMID: 10439033

[40] Cederbrant K, Hultman P, Marcusson JA, Tibbling L. In vitro lymphocyte proliferation as compared to patch test using gold, palladium and nickel. International Archives of Allergy and Immunology. Mar 1997;**112**(3):212-217. DOI: 10.1159/000237456. PMID: 9066505

[41] Valentine-Thon E, Schiawara HW. Validity of MELISA for metal sensitivity testing. Neuroendocrinology Letters. February–April 2003;**24**(1-2):57-64. PMID: 12743534

[42] Stejskal VD, Forsbeck M, Nilsson R. Lymphocyte transformation test for diagnosis of isothiazolinone allergy in man. Journal of Investigative Dermatology. June 1990;**94**(6):798-802. DOI: 10.1111/1523-1747.ep12874656. PMID: 1693940

[43] Willis CM, Young E, Brandon DR, Wilkinson JD. Immunopathological and ultrastructural findings in human allergic and irritant contact dermatitis. British Journal of Dermatology. September 1986;**115**(3):35-16. DOI: 10.1111/j.1365-2133.1986.tb05745.x. PMID: 3530310

[44] Karam J. Methods in Nucleic Acids Research. Boca Raton: CRC Press; November 26, 1990. p. 309. ISBN: 0849353114

[45] Burtis CA, editor. Tietz Textbook of Clinical Biochemistry and Molecular Diagnostics. USA: Elsevier Saunders; 2012. p. 393. ISBN: 9781416061649

[46] Concurrent Engineering for Lateral-Flow Diagnostics (IVDT archive, Nov 99). Archived 2014-04-15 at the Wayback Machine

[47] "Archived copy". Archived from the original on 2012-07-28. Retrieved 2012-07-27

[48] Hansson J, Yasuga H, Haraldsson T, van der Wijngaart W. Synthetic microfluidic paper: High surface area and high porosity polymer micropillar arrays. Lab on a Chip. 2016;**16**:298-304. DOI: 10.1039/C5LC01318F

[49] Guo W, Hansson J, van der Wijngaart W. Viscosity Independent Paper Microfluidic Imbibition" (PDF). MicroTAS 2016; Dublin, Ireland; 2016

[50] Edwards EA, Phillips IA, Sulter WC. Diagnosis of group A streptococcal infections directly from throat secretions. Journal of Clinical Microbiology. 1982;**15**:481-483

[51] Smits HL, Abdoel TH, Solera J, Clavijo E, Diaz R. Immunochromatographic Brucella specific immunoglobulin M and G lateral flow assays for rapid serodiagnosis of human brucellosis. Clinical and Diagnostic Laboratory Immunology. 2003;**10**(6):1141-1146

[52] Nikitin PI, Vetoshko PM, Ksenevich TI. Magnetic immunoassays. Sensor Letters. 2007;**5**: 1-4

[53] Rubenstein R et al. Prion disease detection, PMCA kinetics, and IgG in urine from sheep naturally/experimentally infected with scrapie and deer with preclinical/clinical chronic wasting disease. Journal of Virology. September 2011;**85**(17):9031-9038. DOI: 10.1128/ jvi.05111-11. PMC 3165845. PMID: 21715495. Retrieved 2011-08-21

[54] Tsai C-T, Robinson PV, Spencer CA, Bertozzi CR. Ultrasensitive antibody detection by agglutination-PCR (ADAP). ACS Central Science. 2016;**2**(3):139-114

[55] Morais S, Puchades R, Maquieira Á. Disc-based microarrays: Principles and analytical applications. Analytical and Bioanalytical Chemistry. July 2016;**408**(17):4523-4534. DOI: 10.1007/s00216-016-9423-1. PMID: 26922341

[56] Mills DR, Peterson RL, Spiegelman S. An extracellular Darwinian experiment with a self-duplicating nucleic acid molecule. Proceedings of the National Academy of Sciences of the United States of America. July 1967;**58**(1):217-224. DOI: 10.1073/pnas.58.1.217. PMC 335620. PMID: 5231602

Chapter 5

Cellular Interaction of Human Eukaryotic Elongation Factor 1A Isoforms

Nunzia Migliaccio, Gennaro Sanità,
Immacolata Ruggiero, Nicola M. Martucci,
Carmen Sanges, Emilia Rippa,
Vincenzo Quagliariello, Ferdinando Papale,
Paolo Arcari and Annalisa Lamberti

Additional information is available at the end of the chapter

http://dx.doi.org/10.5772/intechopen.74733

Abstract

Besides its canonical role in protein synthesis, the eukaryotic translation elongation factor 1A (eEF1A) is also involved in many other cellular processes such as cell survival and apoptosis. We showed that eEF1A phosphorylation by C-Raf *in vitro* occurred only in the presence of eEF1A1 and eEF1A2, thus suggesting that both isoforms interacted in cancer cells (heterodimer formation). This hypothesis was recently investigated in COS-7 cells where fluorescent recombinant eEF1A isoforms colocalized at the level of cytoplasm with a FRET signal more intense at plasma membrane level. Here, we addressed our attention in highlighting and confirming this interaction in a different cell line, HEK 293, normally expressing eEF1A1 but lacking the eEF1A2 isoform. To this end, His-tagged eEF1A2 was expressed in HEK 293 cells and found to colocalize with endogenous eEF1A1 in the cytoplasm, also at the level of cellular membranes. Moreover, FRET analysis showed, in this case, the appearance of a stronger signal mainly at the level of the plasma membrane. These results confirmed what was previously observed in COS-7 cells and strongly reinforced the interaction among eEF1A isoforms. Moreover, the formation of eEF1A heterodimer in cancer cells could also be important for cytoskeleton rearrangements rather than for phosphorylation, most likely occurring during cell survival and apoptosis.

Keywords: eukaryotic translation elongation factor 1A (eEF1A), confocal microscopy, FRET, pull-down assay, immunoblotting

1. Introduction

Eukaryotic elongation factor 1A (eEF1A) belongs to the family of GTP-binding proteins and it is the second most abundant protein in the cellular environment. It catalyzes the first step of the elongation cycle by promoting the GTP-dependent binding of aminoacyl-tRNA to the A-site of the ribosome [1–3]. eEF1A exists as two isoforms eEF1A1 and eEF1A2 [4], and in humans, they share almost identical amino acid sequences (92% sequence identity). eEF1A1 is ubiquitously present except in skeletal and cardiac muscle, while eEF1A2 expression is restricted in the brain, skeleton muscle, heart, and other cell types including large motor neurons, islet cells in the pancreas, and neuroendocrine cells in the gut [5], and it is currently found in all vertebrates [6]. Besides their role in polypeptide synthesis, paralogous human eEF1A1 and eEF1A2 act as "moonlighting" proteins [7] owing to several noncanonical functions such as cytoskeleton remodeling by binding and bundling filamentous actin [8, 9], apoptosis, nuclear transport, proteasome-mediated degradation of damaged proteins, heat shock, and transformation [10–12]. Overexpression of eEF1A1 or eEF1A2 in Hela cells led to increased cell growth [7], whereas the disruption of eEF1A1 resulted in actin cytoskeleton defects under basal conditions and in response to palmitate, thus suggesting that eEF1A1 mediates lipotoxic cell death, secondary to oxidative and ER stress, by regulating cytoskeletal changes critical for this process [13]. These findings highlighted that eEF1A1 was involved in both cell proliferation and apoptosis, though the relationship between eEF1A1 and apoptosis is still unclear. By contrast, eEF1A2 seems to play antiapoptotic properties in ovarian, breast, pancreatic, liver, and lung cancer [14]; however, this oncogenic potential deserves further investigation [15].

The possible interaction between eEF1A molecules was first characterized in *Tetrahymena* as eEF1A dimer was able to bundle actin filament [16]. Subsequently, the identification of dimeric eEF1A was also reported in both chicken and human B cell lines [17]. Recent investigations indicated that, compared to eEF1A2, eEF1A1 showed a higher property of self-association [18]. Moreover, under oxidant condition, eEF1A1 was able to form intermolecular disulfide bonds [19]. Recent findings showed that C-Raf kinase interacts *in vivo* with eEF1A during a survival response mediated by epidermal growth factor (EGF) following the treatment of human lung cancer cells with α-interferon (IFNα) [20]. Moreover, phosphorylation of *e*EF1A *in vitro* by C-Raf on S21 required the presence of both *e*EF1A isoforms, thus suggesting that the existence of an eEF1A1/eEF1A2 complex and the S21 phosphorylation represented a regulatory mechanism responsible for the switch from eEF1A canonical to noncanonic functions [21]. On the basis of these findings, we recently showed the possible direct interaction between the *e*EF1A isoforms by using fluorescence resonance energy transfer (FRET) [22]. Compared to our previous work, here we settled for a different experimental approach mainly based on pull-down, confocal microscopy, and FRET analysis based on IgG-FITC (donor)- and IgG-TRITC (acceptor)-conjugated antibodies in HEK 293 cells transfected with recombinant His-tagged eEF1A2 isoform.

2. Expression and interaction of eEF1A1 and eEF1A2 in HEK 293 cells

To assess the possible physiological interaction between eEF1A isoforms in a natural cellular environment, such as the cytoplasm of intact cells, human embryonic kidney 293 (HEK 293) cell line was used as an experimental system. This choice was derived from the finding that HEK 293 cells normally express substantial levels of eEF1A1 isoform, whereas the eEF1A2 isoform is absent.

2.1. Expression in HEK 293 of eEF1A1 and eEF1A2

First, the efficiency of pcDNA3.1-eEF1A2(His)$_6$ (gift from C. R. Knudsen, Aarhus, Denmark [23]) to transfect HEK 293 cells was evaluated. As reported in **Figure 1A**, compared to non-transfected HEK 293, cells transfected with recombinant eEF1A2 isoform showed an increase in the expression of the 54 kDa bands corresponding to the molecular weight of eEF1A. Subsequently, the expression level of eEF1A2 using a specific anti-eEF1A2 antibody (prepared as already reported [22]) was analyzed. As shown in **Figure 1B**, eEF1A2 isoform was revealed only in HEK 293 cells transfected with pcDNA3.1-eEF1A2(His)$_6$ and confirmed with the anti-His antibody (Merck, Germany) (**Figure 1C**).

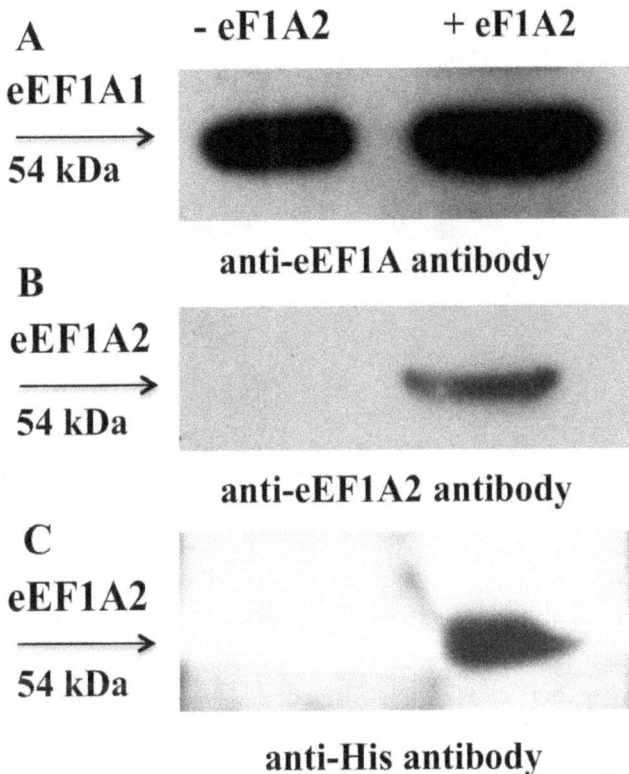

Figure 1. Expression of eEF1A isoforms in HEK 293 cells. HEK 293 cells were transfected with pcDNA3.1-eEF1A2(His)$_6$, and after 24 h from transfection, cell extracts were analyzed by Western blot using commercial mouse anti-eEF1A antibody (A), anti-eEF1A2 antibody (B), and rabbit anti-His antibody (C). Lanes: −eEF1A2, non-transfected HEK 293 cells; +eEF1A2, HEK 293 cells transfected with pcDNA3.1-eEF1A2(His)$_6$.

2.2. Both eEF1A1 and eEF1A2 immuno-interact after pull-down

The possible interaction between eEF1A isoforms was analyzed by pull-down experiment. To this purpose, GST-eEF1A1 (kindly supplied by C. Sanges, Wurzburg, Germany [21]) and pcDNA3.1-eEF1A2(His)$_6$ constructs were co-transfected in HEK 293 cells and, after 24 h from transfection, cell extracts were analyzed by Western blot following GST-agarose and Ni-NTA-agarose pull-down. As shown in **Figure 2**, compared to controls, GST pull-down of co-transfected cells showed the presence of a band of 54 kDa corresponding to the size of eEF1A2(His)$_6$ (**Figure 2A**, lane 2), whereas Ni-NTA pull-down showed the presence of a band of about 78 kDa corresponding to the size of the construct GST-eEF1A1 (**Figure 2B**, lane 2). **Figure 2B** (lane 3) also shows the presence of a band of about 26 kDa corresponding to the GST protein. This finding suggested that GST by itself somehow interacted with Ni-NTA matrix; thus, the result shown in line 2 could be partly due to an interaction of the GST moiety present in GST-eEF1A1 with Ni-NTA and not with eEF1A2. Therefore, to further confirm the interaction between eEF1A isoforms, a different approach was undertaken after transfection of HEK 293 cells with pcDNA3.1-eEF1A2(His)$_6$. In fact, as reported in **Figure 2C**, compared to cells transfected with pcDNA3.1 empty vector, cells transfected with eEF1A2(His)$_6$ showed, after Ni-NTA

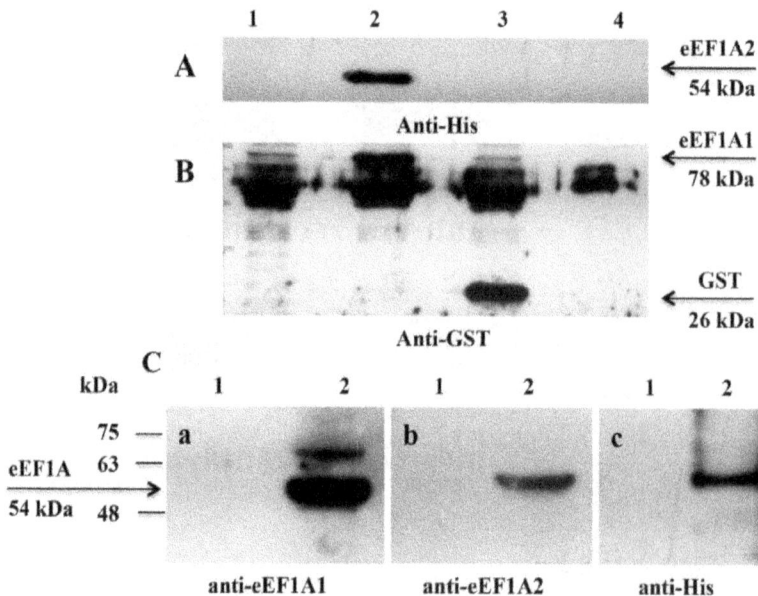

Figure 2. Co-transfection of GST-eEF1A1 and eEF1A2-His in HEK 293 cells. GST-eEF1A1 and pcDNA3.1-eEF1A2(His)$_6$ were cotransfected in HEK 293 7 cells. After 24 h, the cells were harvested, lysed, and analyzed after GST pull-down with antibody anti-His (A) and after Ni-NTA pull-down with anti-GST antibody (B). (A) Lanes: 1, non-transfected cells; 2, cells transfected with GST-eEF1A1 and pcDNA3.1-eEF1A2(His)$_6$; 3, cells transfected with GST and pcDNA3.1-eEF1A2(His)$_6$; 4, GST-agarose alone. (B) Lanes: 1, non-transfected cells; 2, cells transfected with GST-eEF1A1 and pcDNA3.1-eEF1A2(His)$_6$; 3, cells transfected with GST and pcDNA3.1-eEF1A2(His)$_6$; 4, Ni-NTA alone. (C) pcDNA3.1-eEF1A2(His)$_6$ was co-transfected in HEK 293 7 cells. After 24 h, the cells were harvested, lysed, and analyzed after Ni-NTA pull-down with anti-eEF1A1, anti-eEF1A2, and anti-His antibody. (a–c) Lanes: 1, cells transfected with empty vector; 2, cells transfected with pcDNA3.1-eEF1A2(His)$_6$.

pull-down of cell extracts, the presence of a band of 54 kDa that was recognized by the specific anti-eEF1A1 (prepared as already reported [22]) (**Figure 2C**-a, lane 2) and anti-eEF1A2 (**Figure 2C**-b, lane 2) antibodies, the latter confirmed also with anti-His antibody (**Figure 2C**-c, lane 2).

2.3. Both eEF1A1 and eEF1A2 colocalize in HEK 293 cells

The intracellular colocalization of eEF1A1 and eEF1A2 was first analyzed by confocal microscopy. As shown in **Figure 3**, HEK 293 cells after 48 h from transfection with pcDNA3.1-eEF1A2(His)$_6$ construct revealed that both endogenous eEF1A (**Figure 3A**) and transfected eEF1A2(His)$_6$ (**Figure 3B**) shared a cytoplasmic localization. The superimposition of the two panels (merged image, **Figure 3D**) showed that both eEF1A isoforms exhibited a cytoplasmic colocalization with specific signals more intense at the level of the plasma membrane.

2.4. FRET analysis showed that both eEF1A1 and eEF1A2 interact in HEK 293 cells

The interaction between endogenous eEF1A and transfected eEF1A2(His)$_6$ was further investigated by sensitized emission FRET method. FRET effect was performed by confocal microscope

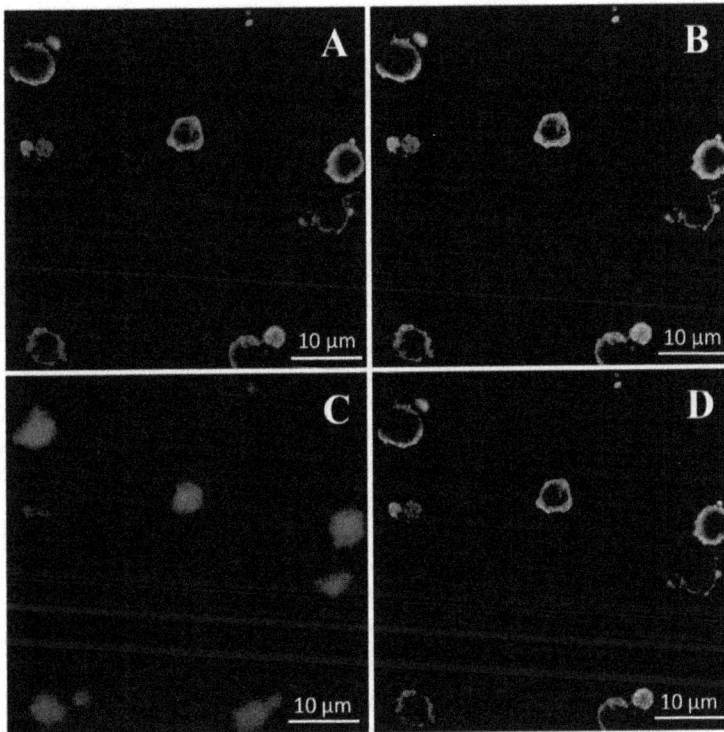

Figure 3. Colocalization of eEF1A1 and eEF1A2(His)$_6$ in HEK 293 cells. HEK 293 cells were transfected with pcDNA3.1-eEF1A2(His)$_6$, and after 48 h from transfection, cells were analyzed by confocal microscopy. (A) eEF1A1, (B) eEF1A2, (C) nuclear staining, and (D) merged images.

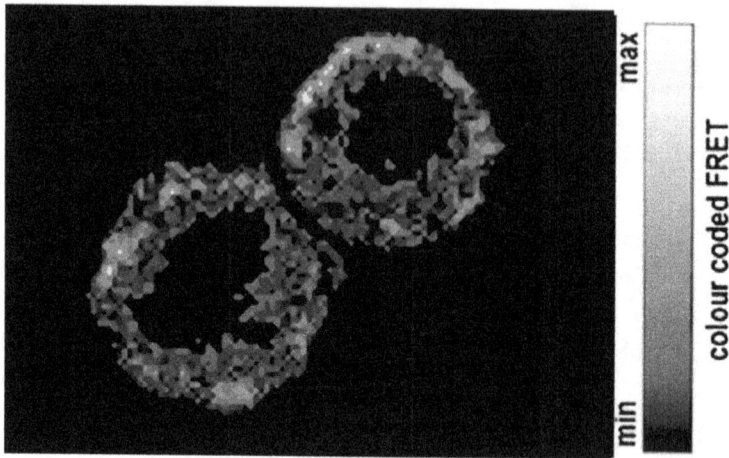

Figure 4. FRET analysis of the interaction between eEF1A1 and eEF1A2. Representative pseudocolor images of cells labeled with rabbit anti-eEF1A1 and mouse anti-His primary antibodies followed by FITC and TRITC secondary-labeled antibodies.

that allowed discriminate proteins that colocalize in the same cellular compartment from those that are instead involved in specific molecular interactions. FRET effects were calculated using ImageJ plug-in software [24]. **Figure 4** shows the representation of the FRET effects where the blue color is indicated at low signal, whereas yellow-white color designated a high signal. The images clearly showed the interaction between eEF1A1 and transfected eEF1A2(His)$_6$ within the cytoplasm with specific signals more intense especially at the level of the plasma membrane.

3. Discussion

FRET is a powerful technique suitable for studying *in situ* interactions between biological molecules in cellular environments [25]. FRET can be assessed from the transfer of energy from one fluorescent molecule (donor) to another fluorescent molecule (acceptor). This process occurs optimally only if the two molecules are properly oriented and reasonably at a narrow distance (usually 1–10 nm) [26]. By this technique, the interaction between eEF1A1 and eEF1A2 in order to reinforce our hypothesis on the formation of an eEF1A1-eEF1A2 heterodimer [21] was highlighted. In a different approach, we afforded this aspect by expressing chimeric eEF1As fused to CFP and YFP as donor and acceptor (CFP-eEF1A1 and YFP-eEF1A2) in COS-7 cells, respectively [22]. However, some criticisms emerged that could have affected the FRET results such as (1) the possible interaction of the expressed chimeric proteins with endogenous enzymes, (2) self-association between eEF1A molecules (i.e., homodimer formation), and (3) the overexpression in COS-7 cells of both constructs that could have generated an art factual FRET signal mainly at the level of plasma membrane. Therefore, to overcome these concerns and to confirm that both eEF1A isoforms interact in the cellular environment, we used a different approach based on the use of IgG-FITC (donor)- and IgG-TRITC (acceptor)-conjugated antibodies. To this end, HEK 293 cell line, lacking the expression of eEF1A2 isoform, was transfected with pcDNA3.1-eEF1A2(His)$_6$, and the interaction of the recombinant eEF1A2(His)$_6$ with endogenous eEF1A1 was assessed by pull-down, confocal microscopy and FRET analysis. The results obtained showed that the endogenous eEF1A1 and the expressed eEF1A2 interacted in HEK 293 cells at the level of both cytoplasm and plasma membrane. Moreover, the FRET image highlighted a more intense signal at the level of the plasma membrane. These data confirmed those reported in our previous work [22], thus strongly confirming the association in the cells of eEF1A isoforms.

The homodimer association of eEF1As has recently emerged from the crystallization of rabbit eEF1A2 [27] or as proposed in *Tetrahymena*, in order to explain actin bundling essential for the regulation of actin cytoskeleton and cell morphology during several cellular processes [16]. The possible association between eEF1A isoforms was instead proposed by Sanges et al. [21] in studying the control of eEF1A function in cancer cells *via* phosphorylation and by Lee et al. [28] in studying the interaction of eEF1A2 with the tumor suppressor protein p16[INK4a]. Since eEF1A1 and eEF1A2 display a very high amino acid sequence identity (above 97%), the overall structures appear quite similar, as can be predicted by bioinformatic analysis at the GRAMM-X docking Web Server v.1.2.0 [29, 30], using rabbit eEF1A2 (PDB 4C0S chain A) as template [27]. These considerations suggest that both eEF1A1 and eEF1A2 complexes are present in the cells either as homodimer or as heterodimer. These complexes are most likely associated with regulatory noncanonical functions of eEF1As.

4. Conclusions

Because eEF1A dimers are involved in actin bundling [31, 32], it emerges that the fraction of eEF1A as dimer is mostly involved in the actin cytoskeleton rearrangement. Therefore, the cellular distribution of eEF1A molecules between monomeric and dimeric form regulates the functional role of eEF1A in translation or in actin bundling. Because actin chains and translational system coexist in the cells and maybe also functionally dependent [33, 34], the transition "monomer-dimer-monomer" of eEF1A should be relatively easy depending on the cell conditions [35]. This interconversion may be regulated by the reversible posttranslational modifications of eEF1A [36] and its interactions with the protein partners such as Raf kinases [20, 21]. Therefore, it is possible that in cells coexpressing both isoforms, like cancer cells, eEF1A heterodimer formation could also be important for cytoskeleton rearrangements rather than for some phosphorylation catalysis most likely occurring during cell survival and apoptosis [20, 21].

5. Materials and methods

5.1. Cell culture and transfection

HEK 293 cells, obtained from the American Type Tissue Collection (Rockville, MD, USA), were grown at 37°C in a 5% CO_2 atmosphere in Dulbecco's modified Eagle medium (DMEM) (Gibco, Monza, Italy) supplemented with 10% heat-inactivated FBS (GIBCO), 100 U/ml penicillin, 100 mg/ml streptomycin, and 1% L-glutamine.

For Western blot analysis, cells (300×10^3/well) were transfected with GST-eEF1A1 (1 μg), pcDNA3.1-eEF1A2(His)$_6$ (1 μg) and pcDNA3.1 (3 μg) as control using Lipofectamine 2000 or K2. Twenty-four hours after transfection, cells were collected and the corresponding extract analyzed with mouse monoclonal anti-eEF1A antibody.

For confocal microscopy and FRET analysis, cells (10×10^3) were layered on 10-mm glass coverslips, grown at confluence and then transfected with pcDNA3.1-eEF1A2(His)$_6$ (1 μg) or with pcDNA3.1 (1 μg) as controls. Cells were analyzed after 24 h of incubation.

5.2. Cytosolic extracts, pull-down assay, and Western blot

After growth, HEK 293 were scraped, washed twice in PBS, resuspended for 30 min on ice in 20–40 μl of lysis buffer (50 mM Tris-HCl pH 7.4, 1% NP40, 0.25% sodium deoxycholate, 150 mM NaCl, 1 mg/ml aprotinin, leupeptin, pepstatin, 1 mM Na3VO4, 1 mM NaF), and then centrifuged at $14,000 \times g$ for 20 min at 4°C.

Pull-down assay for GST-eEF1A1 or eEF1A2(His)$_6$ was carried out using GST-sepharose (Amersham, Milan, Italy) or Ni-NTA agarose (Qiagen, Milan, Italy), respectively. In detail, 500 μg of cell extracts was incubated with pre-equilibrated resin (about 150 μl slurry/1 mg protein extract) for 2 h at room temperature (RT) or ON at 4°C, respectively. Subsequently, for GST pull-down, the resin was washed two times (centrifugation for 2 min at 2000 r.p.m. 4°C) with 1 ml of 1× phosphate-buffered saline (PBS), whereas for Ni-NTA pull-down, the resin was washed two times with 50 mM NaH$_2$PO$_4$, 300 mM NaCl, and 20 mM imidazole, to reduce nonspecific bound proteins, 0.05% Tween 20, pH 8.0. Successively, the samples were resuspended in 30 μl of 4× Laemmli loading buffer, heated to 95°C for 15 min and subjected to Western blot analysis.

Protein concentration was determined by a modified Bradford method, using the Bio-Rad protein assay and compared with bovine serum albumin (BSA) standard curve. Blots were developed using enhanced chemiluminescence detection (SuperSignal West Pico, Pierce, Milan, Italy). All films were scanned using Adobe Photoshop Software (San Jose, CA, USA).

5.3. Confocal laser scanning microscopy

Human embryonic kidney cells (HEK 293 Cell line) were treated for 20 min with glutaraldehyde 2.5% in PBS, washed three times with PBS, permeabilized for 10 min with 0.1% Triton-X100 and finally washed in PBS. Cells were then blocked for 20 min with 1% BSA in PBS, and after apposite washes, cells were incubated with rabbit anti-EF1A1 antibody (GenScript, Piscataway, NJ, USA) and mouse anti-His polyclonal antibody (Santa Cruz Biotechnology, Santa Cruz, CA, USA) diluted 1:300 in 1% BSA for 1 h. After washing three times with PBS, cells were incubated for 1 h with the appropriate secondary antibodies conjugated to fluorochromes and diluted 1:1000 in 1% BSA. Incubation with TOPRO 3-Iodide (Invitrogen Molecular Probes Eugene, OR, USA) diluted 1/1000 in BSA 1% was done for staining of the nucleus. After this, cells were washed properly with PBS and then observed with a Nikon Confocal Microscope C1 equipped with an EZ-C1 Software for data acquisition by using 60× oil immersion objective.

5.4. FRET analysis

HEK 293 cells (7×10^3 cells/cm^2) were grown for 24 h on glass coverslips under standard conditions (37°C, 5% CO_2). Cells were then rinsed with PBS, fixed for 10 min with formaldehyde (3.7% in PBS), permeabilized for 10 min with Triton X-100 (0.1% in PBS), and blocked for 20 min in bovine serum albumin (BSA) (1% in PBS). Subsequently, each sample was incubated for 1 h with 5 μg/ml of mouse anti-His and 5 μg/ml of human anti-eEF1A1 antibodies. Following PBS washes, cells were treated for 1 h with goat anti-mouse IgG FITC-conjugated antibody (donor) (Santa Cruz Biotechnology, Santa Cruz, CA, USA) (1 μg/ml) and with goat anti-rabbit IgG-TRITC-conjugated antibody (acceptor) (Santa Cruz Biotechnology, Santa Cruz, CA, USA) (10 μg/ml), Finally, after 3× washes in PBS, confocal images were collected using a Nikon Confocal Microscope C1 furnished with EZ-C1 software. FRET analysis was carried out as already reported [24]. "FRET" images give the calculated amount of FRET for each pixel in the merged images. The ImageJ plug-in color codes the relative FRET efficiency, which is reported by the displayed color bar, on the right of the images.

Acknowledgements

This work was supported by funds from Programmi di Ricerca Scientifica di Rilevante Interesse Nazionale (2012CK5RPF_004), PON Ricerca e Competitività 2007-2013 (PON01_02782), and POR Campania FSE 2007-2013, Project CRÈME.

Conflicts of interest

The authors declare that there is no conflict of interest.

Disclosure statement

Nothing to declare.

Author contributions

NM, IR, and NMM were involved in WB analysis and cell extract preparation; GS and CS were involved in cell transfection; ER was in charge of tissue culture; VQ and FP performed confocal and FRET analysis; PA and AL were involved in the reading and approval of the manuscript.

Abbreviations

CFP	cyan fluorescent protein
DMEM	Dulbecco's modified Eagle medium
eEF1A	eukaryotic elongation translation factor 1A
EGF	epidermal growth factor
FBS	fetal bovine serum
FRET	fluorescence resonance energy transfer
GST	glutathione S-transferase
Ni-NTA	nickel-nitrilotriacetic acid
PBS	phosphate-buffered saline
TRITC	tetramethylrhodamine
YFP	yellow fluorescent protein

Author details

Nunzia Migliaccio[1], Gennaro Sanità[1], Immacolata Ruggiero[1], Nicola M. Martucci[1], Carmen Sanges[2], Emilia Rippa[1], Vincenzo Quagliariello[3], Ferdinando Papale[4], Paolo Arcari[1,5*] and Annalisa Lamberti[1]

*Address all correspondence to: arcari@unina.it

1 Department of Molecular Medicine and Medical Biotechnology, University of Naples Federico II, Naples, Italy

2 ERT, eResearchTechnology GmbH, Estenfeld, Germany

3 Department of Abdominal Oncology, National Cancer Institute "G. Pascale", Naples, Italy

4 Department of Thoracic and Cardio-respiratory Sciences, University of Campania L. Vanvitelli, Naples, Italy

5 CEINGE, Advanced Biotechnology Scarl, Naples, Italy

References

[1] Kaziro Y, Itoh H, Kozasa T, Nakafuku M, Satoh T. Structure and function of signal-transducing GTP-binding proteins. Annual Review of Biochemistry. 1991;**60**:349-400. DOI: 10.1146/annurev.bi.60.070191.002025

[2]   Klink F. In: Woese CR, Wolfe R, editors. The Bacteria. Vol. 8. London: Academic Press;
      1985. pp. 379-410

[3]   Moldave K. Eukaryotic protein synthesis. Annual Review of Biochemistry. 1985;**54**:
      1109-1149. DOI: 10.1146/annurev.bi.54.070185.005333

[4]   Lund A, Knudsen SM, Vissing H, Clark B, Tommerup N. Assignment of human elon-
      gation factor 1alpha genes: EEF1A maps to chromosome 6q14 and EEF1A2 to 20q13.3.
      Genomics. 1996;**36**:359-361. DOI: 10.1006/geno.1996.0475

[5]   Newbery HJ, Loh DH, O'Donoghue JE, Tomlinson VAL, Chau YY, Boyd JA, Bergmann JH,
      Brownstein D, Abbott CM. Translation elongation factor eEF1A2 is essential for post-
      weaning survival in mice. The Journal of Biological Chemistry. 2007;**282**:28951-28959.
      DOI: 10.1074/jbc.M703962200

[6]   Newbery HJ, Stancheva I, Zimmerman LB, Abbott CM. Evolutionary importance of trans-
      lation elongation factor eEF1A variant switching: EEF1A1 downregulation in muscle is
      conserved in Xenopus but is controlled at a posttranscriptional level. Biochemical and
      Biophysical Research Communications. 2011;**411**:19-24. DOI: 10.1016/j.bbrc.2011.06.062

[7]   Mateyak MK, Kinzy TG. eEF1A: Thinking outside the ribosome. The Journal of Biological
      Chemistry. 2010;**285**:21209-21213. DOI: 10.1074/jbc.R110.113795

[8]   Murray JW, Edmonds BT, Liu G, Condeelis J. Bundling of actin filaments by elonga-
      tion factor 1α inhibits polymerization at filament ends. The Journal of Cell Biology.
      1996;**135**:1309-1321. DOI: 10.1083/jcb.135.5.1309

[9]   Gross SR, Kinzy TG. Translation elongation factor 1A is essential for regulation of
      the actin cytoskeleton and cell morphology. Nature Structural & Molecular Biology.
      2005;**12**:772-778. DOI: 10.1038/nsmb979

[10] Ejiri S. Moonlighting functions of polypeptide elongation factor 1: From actin bundling
      to zinc finger protein R1-associated nuclear localization. Bioscience, Biotechnology,
      and Biochemistry. 2002;**66**:1-21. DOI: 10.1271/bbb.66.1

[11] Chuang SM, Chen L, Lambertson D, Anand M, Kinzy TG, Madura K. Proteasome-
      mediated degradation of cotranslationally damaged proteins involves translation
      elongation factor 1A. Molecular and Cellular Biology. 2005;**25**:403-413. DOI: 10.1128/
      MCB.25.1.403-413.2005

[12] Hotokezaka Y, Tobben U, Hotokezaka H, Van Leyen K, Beatrix B, Smith DH, Nakamura T,
      Wiedmann M. Interaction of the eukaryotic elongation Factor 1A with newly synthe-
      sized polypeptides. The Journal of Biological Chemistry. 2002;**277**:18545-18551. DOI:
      10.1074/jbc.M201022200

[13] Borradaile NM, Buhman KK, Listenberger LL, Magee CJ, Morimoto ET, Ory DS,
      Schaffer JE. A critical role for eukaryotic elongation factor 1A-1 in lipotoxic cell death.
      Molecular Biology of the Cell. 2006;**17**:770-778. DOI: 10.1091/mbc.E05-08-0742

[14] Lee MH, Surh YJ. eEF1A2 as a putative oncogene. Annals of the New York Academy of
      Sciences. 2009;**1171**:87-93. DOI: 10.1111/j.1749-6632.2009.04909.x

[15] Tomlinson VA, Newbery HJ, Wray NR, Jackson J, Larionov A, Miller WR, Dixon JM, Abbott CM. Translation elongation factor eEF1A2 is a potential oncoprotein that is overexpressed in two-thirds of breast tumors. BMC Cancer. 2005;5:113. DOI: 10.1186/1471-2407-5-113

[16] Bunai F, Ando K, Ueno H, Numata O. Tetrahymena eukaryotic translation elongation factor 1A (eEF1A) bundles filamentous actin through dimer formation. Journal of Biochemistry. 2006;140:393-399. DOI: 10.1093/jb/mvj169

[17] HaÅNsler J, Rada C, Neuberger MS. The cytoplasmic AID complex. Seminars in Immunology. 2012;24:273-280. DOI: 10.1016/j.smim.2012.05.004

[18] Timchenko AA, Novosylna OV, Prituzhalov EA, Kihara H, El'skaya AV, Negrutskii BS, Serdyuk IN. Different oligomeric properties and stability of highly homologous A1 and proto-oncogenic A2 variants of mammalian translation elongation factor eEF1. Biochemistry. 2013;52:5345-5353. DOI: 10.1021/bi400400r

[19] Cumming RC, Andon NL, Haynes PA, Park M, Fischer WH, Schubert D. Protein disulfide bond formation in the cytoplasm during oxidative stress. The Journal of Biological Chemistry. 2004;279:21749-21758. DOI: 10.1074/jbc.M312267200

[20] Lamberti A, Longo O, Marra M, Tagliaferri P, Bismuto E, Fiengo A, Viscomi C, Budillon A, Rapp UR, Wang E, Venuta S, Abbruzzese A, Arcari P, Caraglia M. C-RAF antagonizes apoptosis induced by IFN-alpha in human lung cancer cells by phosphorylation and increase of the intracellular content of elongation factor 1A. Cell Death and Differentiation. 2007;14:952-962. DOI: 10.1038/sj.cdd.4402102

[21] Sanges C, Scheuermann C, Zahedi RP, Sickmann A, Lamberti A, Migliaccio N, Baljuls A, Marra M, Zappavigna S, Rapp U, Abbruzzese A, Caraglia M, Arcari P. Raf kinases mediate the phosphorylation of eukaryotic translation elongation factor 1A and regulate its stability in eukaryotic cells. Cell Death & Disease. 2012;3:e276. DOI: 10.1038/cddis.2012.16

[22] Migliaccio N, Ruggiero I, Martucci NM, Sanges C, Arbucci S, Tatè R, Rippa E, Arcari P, Lamberti A. New insights on the interaction between the isoforms 1 and 2 of human translation elongation factor 1A. Biochimie. 2015;118:1-7. DOI: 10.1016/j.biochi.2015.07.021

[23] Kahns S, Lund A, Kristensen P, Knudsen CR, Clark BF, Cavallius J, Merrick WC. The elongation factor 1 A^{-2} isoform from rabbit: Cloning of the cDNA and characterization of the protein. Nucleic Acids Research. 1998;26:1884-1890. DOI: 10.1093/nar/26.8.1884

[24] Hachet-Haas M, Converset N, Marchal O, Matthes H, Gioria S, Galzi JL, Lecat S. FRET and colocalization analyzer method to validate measurement of sensitized emission FRET acquired by confocal microscopy and available as an ImageJ plug-in. Microscopy Research and Technique. 2006;69:e941-e956. DOI: 10.1002/jemt.20376

[25] Sekar RB, Periasamy A. Fluorescence resonance energy transfer (FRET) microscopy imaging of live cell protein localizations. The Journal of Cell Biology. 2003;160:629-633. DOI: 10.1083/jcb.200210140

[26] Konig P, Krasteva G, Tag C, Konig IR, Arens C, Kummer W. FRET-CLSM and double-labeling indirect immunofluorescence to detect close association of proteins in tissue sections. Laboratory Investigation. 2006;**86**:853-864. DOI: 10.1038/labinvest.3700443

[27] Crepin T, Shalak VF, Yaremchuk AD, Vlasenko DO, McCarthy A, Negrutskii BS, Tukalo MA, El'skaya AV. Mammalian translation elongation factor eEF1A2:X-ray structure and new features of GDP/GTP exchange mechanism in higher eukaryotes. Nucleic Acids Research. 2014;**42**:12939-12948. DOI: 10.1093/nar/gku974

[28] Lee MH, Choi BY, Cho YY, Lee SY, Huang Z, Kundu JK, Kim MO, Kim DJ, Bode AM, Surh YJ, Dong Z. Tumor suppressor p16INK4a inhibits cancer cell growth by downregulating eEF1A2 through a direct interaction. Journal of Cell Science. 2013;**126**:3796. DOI: 10.1242/jcs.113613

[29] Tovchigrechko A, Vakser IA. Development and testing of an automated approach to protein docking. Proteins. 2005;**60**:296-301. DOI: 10.1002/prot.20573

[30] Tovchigrechko A, Vakser IA. GRAMM-X public web server for protein-protein docking. Nucleic Acids Research. 2006;**34**:W310-W314. DOI: 10.1093/nar/gkl206

[31] Stevenson RP, Veltman D, Machesky LM. Actin-bundling proteins in cancer progression at a glance. Journal of Cell Science. 2012;**125**:1073-1079. DOI: 10.1242/jcs.093799

[32] Tolbert CE, Burridge K, Campbell SL. Vinculin regulation of F-actin bundle formation: What does it mean for the cell? Cell Adhesion & Migration. 2013;**7**:219-225. DOI: 10.4161/cam.23184

[33] Stapulionis R, Kolli S, Deutscher MP. Efficient mammalian protein synthesis requires an intact F-actin system. The Journal of Biological Chemistry. 1997;**272**:24980-24986. DOI: 10.1074/jbc.272.40.24980

[34] Perez WB, Kinzy TG. Translation elongation factor 1A mutants with altered actin bundling activity show reduced aminoacyl-tRNA binding and alter initiation via eIF2alpha phosphorylation. The Journal of Biological Chemistry. 2014;**289**:20928-20938. DOI: 10.1074/jbc.M114.570077

[35] Vlasenko DO, Novosylna OV, Negrutskii BS, El'skaya AV. Truncation of the A,A/,A0 helices segment impairs the actin bundling activity of mammalian eEF1A1. FEBS Letters. 2015;**589**:1187-1193. DOI: 10.1016/j.febslet.2015.03.030

[36] Negrutskii B, Vlasenko D, El'skaya A. From global phosphoproteomics to individual proteins: The case of translation elongation factor eEF1A. Expert Review of Proteomics. 2012;**9**:71-83. DOI: 10.1586/epr.11.71